The Death of Annie

C. Thomas Cook, M.D.

ISBN 978-1-64670-314-2 (Paperback)
ISBN 978-1-64670-315-9 (Digital)

Copyright © 2020 C. Thomas Cook, M.D.
All rights reserved
First Edition

All rights reserved. No part of this publication may be reproduced, distributed, or transmitted in any form or by any means, including photocopying, recording, or other electronic or mechanical methods without the prior written permission of the publisher. For permission requests, solicit the publisher via the address below.

Covenant Books, Inc.
11661 Hwy 707
Murrells Inlet, SC 29576
www.covenantbooks.com

Contents

Acknowledgments 5
The Death of Annie 7

Volume One

Chapter 1: Introduction 13
Chapter 2: The Case for the Transcendent 18
Chapter 3: Perspective 25
Chapter 4: The Human Brain 35
Chapter 5: Science and Darwin's Theory 50
Chapter 6: General Thoughts: Science, Teaching, and the Transcendent 55
Chapter 7: Chance and Probability 59
Chapter 8: Random 68
Chapter 9: Determinism 71
Chapter 10: Consciousness 74
Chapter 11: The Theory of Evolution 78
Chapter 12: Strengths, Weaknesses, and Assumptions 88
Chapter 13: Science 97
Chapter 14: Artificial Intelligence 110
Chapter 15: Analysis 115
Chapter 16: Summary 120
Chapter 17: Epilogue 126

Acknowledgments

I would like to thank Rev. Deane Kemper, the first to read my manuscript and who encouraged me to continue its writing and refinement. Dr. Robert Ogilvie, who reviewed the manuscripts and offered many insights and editorial comments regarding the scientific, philosophical, and spiritual content. To the Harborview Presbyterian Church Book Club, especially its leader, Jan Kucklick. To my wife, Sandra, who listened to and read my many comments and revisions and gave her constructive advice and encouragement. To Reverent Rusty Benton for his enthusiastic support and helpful input and sentiments. And finally, to the Outside Force, without whose help this book would have been impossible.

The Death of Annie

My daughter found Annie in an animal shelter in the summer of 1995. She was a small dog, about four or five pounds, with short, coarse, somewhat frizzled, black-and-white hair. Not much to look at, but with sad eyes that spoke of a melancholy past, which captured the emotions of even the most casual observer.

Annie's eyes would stay that way even after three years of care and love by her adopted family. The initial lethargy and confusion would respond to good food and attention. In time, Annie became playful and active just as any middle-aged dog would do, but the sadness never left. Just as with people, traumas heal but leave indelible scars. I never caught her staring out of the window, but had she been taller, no doubt there would have been those times, like all of us, when there's nothing else to do but to look back, regardless of what we may see. Nostalgia—painful remembrance.

When Annie became ill, and the veterinarian told us that she had a severe form of diabetes, our reaction was acceptance and resignation, mingled with what could be called a sense of expectation.

The next year was filled with the remissions and exacerbations that are part of the diabetic condition. Infections, vomiting, weight loss, pain, lethargy…good times and bad. It's hard to know if a dog ever complains—I've never seen one do so. If Annie ever did, it was well masked.

When the end was near and everyone realized it, the younger children, as nature has provided for them, went into their detached mode. The learned call this denial, but anyone who has been through these ordeals with children knows that what's at play is not so shallow as that overused and misused diagnosis but something much more profound than that, which we adults, now older and wiser, have now become incapable of understanding. Too bad. What a gift to retain.

THE DEATH OF ANNIE

The slow ebbing away of Annie's life was too much for my daughter to bear. That's when, the inevitable just hours away, Annie was brought to my house.

Five or six years ago, two stray mixed-breed dogs had a chance rendezvous, and out of this Annie came into the world. No great event. Just another happening. Nothing new. It happens all the time. Just another life. Nursed by the mother. Part of a litter. Out on her own. Picked up. Adopted. Abused. Neglected. Deserted. In a shelter. Adopted by a good family. Terminal diabetes. Short story. Just another dog's life.

Only a matter of hours now. Still breathing. Heart still beats. Some movement of the head. Occasional stretching of the limbs. Eyes drift open and slowly close again. Respirations stop. Faint pulsations noted through the chest wall. Heart stops. Life ends.

What's going on? We are want to say life is *oozing away*. But what is going on here is clearly beyond our ability to comprehend. Maybe we should not try to comprehend things that we cannot comprehend. It's better left that way, rather than confuse ourselves and lose out on the experience of that greatest of all mysteries. I watched as the breathing became slower and slower and stopped. Then the faint movement of the chest wall stopped as the little heart stopped. One last spasm. And it all was over. No more "life."

But the physical elements remain exactly the same. What's the difference between one minute before "death" and one minute after "death"? Is it just biochemical? Is it some "force" that we can't measure? Comprehend? The Great Mystery?

As a physician, I have witnessed many times over the last minute, the last seconds, of human life ebb away. I lived through weeks of my mother on life support and with no apparent awareness of anything. But in all these instances, the difference in our perceptions before and immediately after the moment of death is incomprehensible. We simply do not know what to think. We are incapable of understanding the difference between "life," with low blood pressure and heartbeat barely audible and respirations shallow and no apparent consciousness, and the status a few seconds later and the body is not "alive."

Some would say, "It's over," or "It's finished." Just another entity responding to the laws of thermodynamics: disorganization, disintegration, entropy. Dust returning to dust.

Or is it over? Is it the end? Period… How dreadful it would be if this is all there is: a series of changes and mutations evolving over millions of years, culminating in a temporary "life," and then casual, meaningless disintegration.

I am reminded of the words of Carl Sagan:

> *The essence of life is not so much the atoms and simple molecules that make us up as the way in which they are put together. Every now and then we read that the chemicals which constitute the human body cost ninety-seven cents or ten dollars or some such figure… We are made mostly of water, which costs almost nothing; the carbon is costed in the form of coal; the calcium in our bones as chalk; the nitrogen in our proteins as air (cheap also); the iron in our blood as rusty nails. If we did not know better, we might be tempted to take all the atoms that make us up, mix them together in a big container and stir. We can do this as much as we want. But in the end all we have is a tedious mixture of atoms.*

This little dog's life. Our lives. Chance? Random mutations with no meaning? "A tedious mixture of atoms"?

With gargantuan understatement, I simply say: I don't think so. This book is my effort to be present *Homo sapiens* as the crowning result of a specific process conceived and motivated by a force that we cannot understand, but in every way is real and present in our lives.

Volume One

Chapter One

Introduction

Condensed: This book is my effort to present a logical and scientific case for the transcendent, that is, those things that transcend proof. I will describe what I mean by the transcendent in the first part of the book and build my case chapter by chapter. The reader should keep in mind that the book itself is like a thread composed of three specific strands. One strand is science, one strand is the theory of evolution, and the third strand is the limitations of the human brain. No one can have any feelings about a transcendent force without considering our origin. In our day and time, no one can study the origins of human beings without giving due thought to Charles Darwin's theories regarding human evolution. This study must include a basic understanding of science. These two strands are dependent upon the third strand, the human brain.

As I take pen in hand—or, rather, keyboard and Word processor—I am acutely aware of the observation made by H. L. Mencken:

> *An author, like any other so-called artist, is a man in whom the normal vanity of all men is so vastly exaggerated that he finds it a sheer impossibility to hold it in. His overpowering impulse is to gyrate before his fellow men, flapping his wings and emitting defiant yells. This being forbidden by the police of all civilized countries, he takes it out by putting his yells on paper. Such is the thing called self-expression.*

THE DEATH OF ANNIE

And so, with due respect to Mencken, I succumb to my overpowering impulse to gyrate and flap my wings, and I put my small oar in the ocean.

*****

It behooves us to enlarge our horizons without fear of attacking the biggest problem, without the pretension of solving them during our brief existence, but with the ambition to leave behind us a little less obscurity than we found. (Lecompte du Noüy)

For centuries, human beings have progressively increased their knowledge and understanding of the physical world in which we live. But our understanding of the transcendent,* those things that cannot be explained, has slowly evolved and changed little from the time of the early Greek philosophers. We stand in awe of those mysteries and emotions that uplift us and add to our lives, always seeking understanding and enlightenment.

From the beginning of recorded history, human beings have had a continuous discussion and debate about the relationship between science and religion—e.g., the real or presumed conflict between objective and nonobjective reasoning and analysis. From the very beginning through the present day, there have been conflicts, tensions, agreements and disagreements, and strong feelings on both sides of the issues.

When asked the question "What is the conflict between science and religion?" I answer by saying, "I cannot answer because I don't understand the question." When looking for answers and understanding, human beings in almost every instance use both the logical and rational abilities of their brains along with the emotional or subconscious qualities of their brains—that is, in addition to thinking

---

* *Transcendent* (*Merriam-Webster*): "Extending or lying beyond the limits of ordinary experience. Being beyond the limits of all possible experience and knowledge. Being beyond comprehension."

INTRODUCTION

logically, we use faith, dreams, past memories, hopes, and common sense to solve the unknown. The poet and the philosopher use inductive *and* logical methods in creating and analyzing their thoughts in much the same way the scientist does.

The scientist gets misled thinking that he can understand the entire natural world using only theories, experimentation, and analysis. The religious gets misled by withdrawing from the natural world, believing he or she can gain wisdom and understanding by reducing their dependence on materialism to a minimum. It seems that in the times in which we live we have been misled from the beginning about their being a conflict between the two approaches to the truth, and we continue through life compartmentalizing these two things that do not lend themselves to being placed in singular and clean-cut compartments. This situation can be compared to water, H_2O. Water is a solid, a liquid, or a gas depending on the circumstances—that is, how much heat is present. In the same way, objective reasoning and subjective reasoning are just part of a continuum: different characteristics depending on circumstances in which they occur.

There are numerous examples that we could use to present and study this conflict. This treatise uses Charles Darwin's theory of the evolution of human beings as an excellent scientific theory to examine, as this theory is one of the basics in the teaching of science to young and old alike. It is a classic example of using the *scientific method*: observing a phenomenon; developing a theory; doing field studies or laboratory experimentation to support or refute the theory; and then presenting the study, with its conclusions to the scientific community and to the public.

Darwin's theory has stood the test of time in the scientific community and with most of the lay public. It also presents a stark challenge to many strongly held beliefs about the creation of the planet Earth and its inhabitants. It intertwines the scientific and the religious as no other scientific work does. It is required teaching in all public schools, and therefore anyone with a basic education has been exposed to the theory. And it affects the sense of the value of human beings to anyone familiar with its basic concepts. The reader should keep in mind that this treatise is intended to present a different point of view of the origins of human beings, their evolution and development, their limitations, and their future. It could be looked at as a third way of examining evolution. All the chapters can be viewed as developing challenges to Darwin's theory and supporting the author's contention.

It is clear to anyone who studies the physical world that each gain in knowledge always leads to more questions, and while many human beings believe we will never reach an endpoint, science, almost by definition, is a quest to find the ultimate logical answer. Ideas, books, and articles come and go, and the new generations rarely read the materials of the distant past. Since the advent of the printing press, such massive amounts of written material has been produced that no human being can begin even to touch the surface, let alone study in detail. In our time, the Internet has caused an explosion of written and visual material. New books are constantly being written, and the present generations are exposed primarily to the new, with minimal exposure to the past, great works of literature or science.

Throughout this book, I will make a big issue that nothing happens without a reason. I emphasize that this is true of everything, including any changes noted in the natural world. I make a big issue of the fact that the human brain is limited and not capable of perceiving and understanding all the information presented to it through its senses. The word *transcendent* and the phrase "Outside Force" are defined and used frequently throughout this treatise. They refer to all those things beyond our ability to define objectively. I assume the possibility of an Outside Force affecting what humans call the natural world and that we can study this possibility objectively.

I am not challenging, nor am I capable of challenging, the generally accepted theories and teachings of evolution, but I am simply trying to make the point that there is another way to look at the *causes* of the evolution of the human species other than the two commonly referred to options:

- *Darwin's theory*, based on changes due to random mutations and survival of the fittest, and

INTRODUCTION

- *Creationism*, the creation of the planet Earth and its life-forms in a short period of time in the relatively recent past.

Almost every discussion of human evolution in our era revolves around these two points of view, with the implied idea that there are no other approaches.

The scientific approach to discovery and solving problems has helped mankind make tremendous improvements in his well-being and comfort and enjoyment of life. We can all agree that its positives outweigh any negatives. But the scientific approach is inadequate to discover all the truths that human beings seek. Total objectivity can sometimes not only be inadequate but paradoxically can lead away from the truth.

I make the case that transcendent things can be true—that they need to be taught, not ignored, or worse, fought. That our overall commitment to, even obsession with, material gains through science is hurting human advancement by refusing to seek the overall truth regarding transcendent possibilities. I include references taken from the great scientists in human history (note, scientists, not philosophers or theologians) to support this claim. Their descriptions of the human's place in our universe is uplifting and inspiring and give each of us renewed understanding of the dignity of every human being and encourages us to live accordingly.

Chapter Two

The Case for the Transcendent

Condensed: I present a basic description of how science—that is, purely objective reasoning, develops its conclusions. One of the most important conclusions reached by modern science is to support the theory of evolution. The generally accepted view of the scientific community is that evolution was a result of chance and probabilities. I emphasize the point that the true scientist ends his study at this point and makes no effort to establish a further or a primary cause. This chapter makes the case that there is no conflict between this method and the inborn desire for human beings to search for that primary cause. That there is something that transcends our ability to completely define or understand but is certainly there and available for us to approach, learn about, and respond to. And we should always keep in mind that in their search for truth, the religious people use math, probabilities, and the general rules of scientific thinking, just as the scientists employ that basic religious proposition, faith. I think that everyone intuitively understands the necessity of this combination of developing an understanding of the world in which we live.

*****

  Darwin's theory of evolution and meticulous documentation as presented in his two works, *The Origin of Species* and *The Descent of* Man, are still considered two of the greatest scientific treatises ever created. By their very nature, they address the *creation* of human beings on the planet. From its publication to the present day, it has forced generations of human beings to include an analysis of Darwin's theories in their studies of values, philosophical questions, spiritual questions, and the meaning of the existence of human beings on the planet Earth. Many scientific advancements challenge our understanding of our place and destiny in the universe. Therefore, all human beings who seek to understand their place in the overall scheme of things need to have a basic grasp of scientific methods, conclusions, and limitations.

From the expansion of our knowledge out into the cosmos to the expansion of our knowledge of the infinitely small subatomic world, our philosophies and our religions must keep pace. There is nothing new about these challenges. When the scientists made it clear that the earth was not flat, that the earth along with the other planets circulated the sun, that the solar system was part of a galaxy, that this galaxy was part of a larger universe, that the universe responded to observable and understandable natural laws, that these laws were relative, plus a host of macro- and micro-observations, the philosophers and theologians adapted to these changes, and their understanding of transcendental truths grew and matured.

In our day and time, the scientific world, with its incredible new technologies and methods, along with the modern methods of presenting information, is constantly confronting the average and nonscientific person with a barrage of new and disturbing ideas. It seems overwhelming, and we are tempted to either withdraw from this world and live the cloistered life or to jump in headfirst and go with the flow of this new world. Very few will choose the cloistered life. So the majority of us who feel the tensions created by this new world must make a conscious effort to do our part in making sure that the transcendent impulses are not minimized or destroyed, and we enter a new, dark age.

I believe that it is necessary to develop a consensus and understanding of what "science" is and is not, and for all of us to understand that the scientific method is not infallible. The history of scientific progress is filled with examples of theories and "laws" that later proved to be wrong and were corrected or modified. Almost everyone is familiar with the history of medicine, with its accepted practices changing and evolving, as new evidence and knowledge became available.

Another concern is the fact that much of scientific research is funded by governments with a specific goal in sight. Many of these goals are related to the military, whether for defense or aggression. Obviously, neither values nor long-term effects are considered. The most spectacular of all these types of projects was the development of the nuclear bomb. The modern Internet with its tentacles in every area of life and existing in almost every society on the planet was developed and moves forward without any consideration of values. Digitalized movies and computer games, with their violence and lack of anything resembling reality, have distorted our very sense of what is real and what is make-believe. And, of course, the multifaceted moral, ethical, and financial problems with the choices created by science in the field of medical treatment.

Most of these things are produced by free societies, and no one would suggest that these things should be censored or completely controlled by political bodies. What sort of controls and restrictions that should be developed is one of the great challenges of our era. There are many examples which we could use to present our case, but I think that Charles Darwin's theory of evolution may be the best example to study, understand, and challenge, because most laypeople have a basic understanding of the theory and the mixed feelings it produces. It is an example that can be used to help laypeople understand the methods, challenges, and limitations of science. It brings together studies, methods, and theories from the nineteenth century through the present day, and should help the layman understand the basic scientific approach. Darwin's theory of evolution has been presented in so many different ways that in our present usage, it is referred to as Darwinism, an *"-ism."* That is, a distinctive doctrine, theory, system, or practice. Any theory or study of human evolution intertwines itself with values.

I urge the reader to keep in mind that I am using the theory of evolution as just an example of the methods that we human beings have developed to study and learn about the universe in which we find ourselves. The explosion of advancements in space exploration, pictures of distant galaxies, overload of multiple methods of interpersonal and international communications, the ability to destroy all that human beings have created over the centuries. The list goes on and on, and we have reached a point where we can no longer refuse to include a study of values in our scientific approaches.

Dealing with an all-inclusive subject such as this requires that the writer and the reader have a rudimentary and sometimes more advanced understanding of the science involved, and also an understanding of philosophy and religion. Our goal should be, for ourselves and for the teaching of students, to present the facts from all points of view and develop a synthesis, realizing the limitations we have and that the study of human evolution is a lifetime study and grows and matures with their understanding of the natural world and transcendental issues.

The general approach of this work will be to present a somewhat simplified and overall description of the history of the evolution of the ideas about the development of the universe as we understand it, beginning with the big bang through our present day, and including the evolution and development of the modern human being, specifically the human brain, the most advanced structure so far at the end of the evolutionary chain. I hope to bring all these ideas together, in an understandable way, to the general public, as we all grapple with this important subject. After presenting the basic ideas, I will present my synthesis by bringing together these ideas into one hopefully understandable and reasonable picture compatible with the science and the transcendental ideas presented. In Volume 2, I include scientific, philosophical, religious articles and writings that support or challenge my conclusions. These articles range from the basic and easily understandable to some of the most complex ideas and studies of the natural world. I have added my abstract to some of the articles. To better understand this book, I urge the reader to read the articles.

My goal is to present the case that there is another way of looking at the evolution of the human species other than the two most popular and well-known: *Darwin's theory* (random mutations and survival of the fittest) and *creationism* (the creation of the planet Earth and its life-forms in a short period of time in the relatively recent past). I believe the study of evolution and specifically Darwin's theory is one of the most important topics that we can study, and we must commit ourselves to presenting the total picture to the upcoming generations because of the inevitable influence it has on our sense of values, specifically our worth as human beings and our place in the universe. Most scientific endeavors are neutral regarding values, but the teaching of evolution certainly involve values, just as many other technical advances, such as the development of atomic energy and the Internet, do.

The belief that there is more to our existence than we can prove objectively is a belief held by a wide majority of human beings throughout history. This is a position held by many of our greatest scientists, and this position can stand up to the challenge of any vigorous objective study of the natural world. My point is not to suggest or recommend any specific religion or philosophy or way of life but to argue that all our scientific and objective studies, after adding so much to the material well-being and comfort of human life, always reach a point where they cannot explain certain phenomena. The scientist can always point out that human's inability to explain certain phenomenon will be explained by further study and the accumulation of knowledge—that is, science will "fill in the gaps." One point of view is that humans will eventually explain everything. The other point of view is that humans will never explain everything because the natural evolution of knowledge always reveals to us how little we know and that there is always "something else."

The large majority of human beings throughout history have never been completely satisfied that we understand everything and that systems of belief inevitably are a part of human existence. In other words, science cannot prove that there is no existence outside of the natural world that humans observe. Not being able to prove something does not mean that it does not exist. We cannot prove a transcendental existence, but this does not mean that it does not exist. My point—this is an issue that is a matter of faith that cannot be proved or disproved by either side and, therefore, is an essential part of any human's education. If our educational institutions do not teach this essential point, intentionally or unintentionally, they create biases in the minds of the students. *In this sense, they are unscientific.*

My overriding objective is to make the case that transcendent things can be true—that they need to be taught, not ignored, or worse, fought against, as they sometimes are. By its very nature, the scientific method is limited to a study of the physical world. But a *laissez-faire* approach—that is, having no restrictions on what this method can produce threatens the very survival of the species, even survival of the planet Earth. The theoretical examples are numerous, and fortunately, the scientific community and humanity at large have blunted most of the horrific efforts to take us down these paths. So it is imperative that the individual scientist have a sense of values and restrictions. When any area of human endeavor is taken to an extreme or wavers away from moral and ethical restrictions, it will inevitably fail or be corrected, just as the religions of the world experience when their followers stray too far away from the teachings of their founders.

We cannot be in a position of teaching or requiring certain scientific subjects, without making sure that the students understand the limitations of science and understand that the scientific method is not the final answer to every subject. We must provide a counterbalance, or in some way help students understand the meaning or value of these subjects. Some will object that teaching values is teaching religion and has no place in public education. But we cannot deny that there are many, many great truths that are a part of human history that are not specifically scientific or proven in the usual sense. To deny that they exist seems to me to be the opposite of education.

To teach Darwin's theory of evolution without offering any alternatives would be like teaching determinism in philosophy without offering any alternatives or other philosophies. Philosophy is theoretical and presents almost all the known alternatives to its students. Teaching Darwin's theory of evolution without offering alternatives appears to me to be close to teaching a value system, which appears to me to be close to teaching religion. Our goal in education should be to help our students learn, or at least be exposed to, all the great ideas that the human race has accumulated over the centuries. A proper education must at least expose the learner, as well as stimulate an interest in, all the ideas that have persisted and been a major part of human life on this planet. A little learning is a dangerous thing is a truism. Inadequate teaching presented as adequate teaching also is a dangerous thing.

And a word about faith and reason. Most people believe in the accepted scientific theories and explanations. Most people believe in the basic theory of evolution and what the scientists say about the fossils without ever actually seeing or studying a fossil. This, of course, is taken by faith that the studies were done properly, truthfully written and presented, and pose no problem. We say the continuing consistency and workability of a theory is proof that the theory itself is true. A better example would be the atomic/molecular theory. This theory answers almost all scientific questions and experiments using this theory have a predictable outcome. The fact that it works "proves" it was right. The scientist pressed the button, and the atomic bomb went off. The prevalent way of teaching Darwin's theory of evolution in our public institutions by not having an answer to so many questions and weaknesses fails this test.

Chapter Three
Perspective

Condensed: This chapter describes the place of human beings on the planet Earth in the overall scheme of the universe. There is a massive amount of "stuff" out there beyond us and a massive amount of "stuff" beneath us, and we live in between these two. We have developed instruments and experiments that have helped us gain tremendous insights into the larger and smaller worlds in which we find ourselves. An understanding of where we reside is essential to understanding from whence we came and where we are headed.

*****

*H*uman beings exist between two worlds: the world of the macro and the world of the micro. We walk the surface of the planet Earth, looking outward with our telescopes and inward with our microscopes, studying and learning about realities that we cannot see with our unaided visual capacities.

The Micro World

The most common microscope (and the first to be invented) is the optical microscope, which uses light to pass through a sample to produce an image. Other major types of microscopes are the fluorescence microscope, the electron microscope (both the transmission electron microscope and the scanning electron microscope), and the various types of scanning probe microscopes.

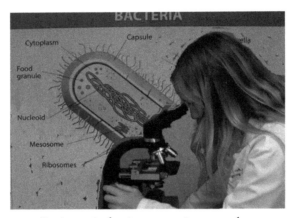

Basic optical microscope in use today.

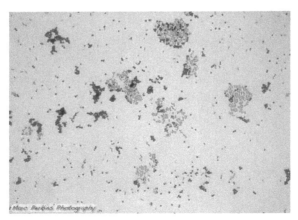

Bacteria, 1000 to 1 magnification, using a basic optical microscope.

PERSPECTIVE

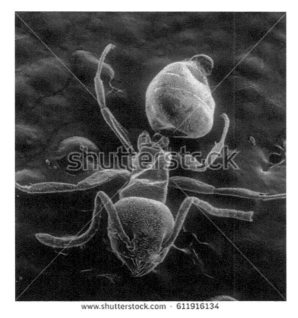

Magnified image of an ant. Image by electron microscope, 2500 magnification.

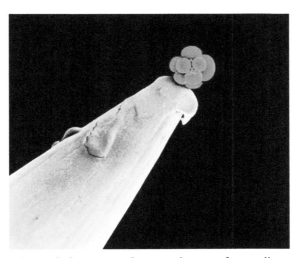

An early human embryo at the tip of a needle.

THE DEATH OF ANNIE

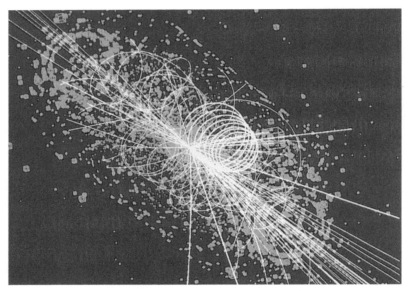

Higgs boson, the smallest bit of matter yet discovered.

PERSPECTIVE

The Macro World

In a universe so large and of such massive proportions of galaxies, stars, and empty space that we simply stand in awe and bewilderment as we try to comprehend the discoveries made by the Hubble telescope and other instruments looking out into space from the planet Earth. As new discoveries are made, we are further struck by realizing we are aware of only a limited portion of the universe, and our limited ability to understand.

The following is a photograph of the night sky as seen with the unaided human eye on a clear night from the planet Earth. Following that are just a few reproductions of the hundreds of pictures taken by the Hubble telescope and NASA moon explorations, but they give us some idea of the place of the planet Earth and us as its inhabitants in the universe.

THE DEATH OF ANNIE

PERSPECTIVE

This is the first image of Earth made from the surface another planet. It was taken by the Mars Exploration Rover *Spirit* on March 8, 2004, an hour before sunrise, with the surface of Mars in the foreground. The contrast was doubled to make Earth easier to see. (*Credit: NASA/JPL/Cornell/Texas A&M*)

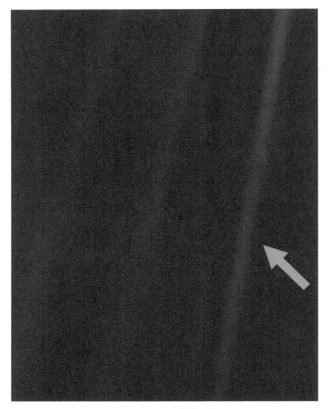

The Pale Blue Dot image is a part of the first portrait ever made of our solar system from afar, taken by *Voyager 1*. *Voyager* acquired 60 frames for a mosaic of the solar system from more than 4 billion miles from Earth and about 32 degrees above the ecliptic—the plane in which most of the planets orbit. Earth is just below and right of center, smack in the middle of one of the scattered light rays resulting from taking the image so close to the sun. (*Credit: NASA*)

PERSPECTIVE

There are billions and billions of stars, more stars than there are grains of sand on all the beaches of the planet Earth. (Attributed to Carl Sagan)

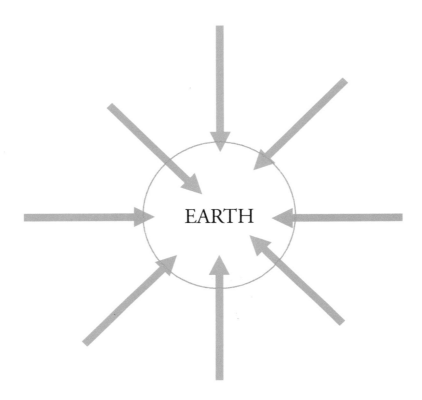

The planet Earth is constantly being bathed in zillions and zillions of various waves: light, sound, x-ray, infrared…and certainly many more that we do not know about. The naked human eye "sees" some of these waves. The human eye sees more with the use of a telescope and with the help of computer-generated pictures from telescopes, and now, with the Hubble telescope. Our knowledge and understanding of the universe in which we live are constantly growing, and so is our awareness of how infinitesimally small we are in the scheme of the universe.

Chapter Four

The Human Brain

I think therefore I am.
—René Descartes

Condensed: This chapter describes the human brain, its abilities, and its limitations. All our knowledge and understanding depends on the abilities of our brains. An awareness of its limitations is essential in understanding human's place in the overall scheme. The human brain is the crown prince of the evolution of our world as we understand it. To fully understand the science of human development and evolution, we must understand the instrument we are using to think about and analyze our observations. This chapter is somewhat long and detailed, but it is offered to enforce the point that the human brain is so complex that the chances of it being a product of randomness exceed the expected standards of scientific analysis.

*****

The brain is an organ that serves as the center of the nervous system in all vertebrate and most invertebrate animals. The brain is located in the head, usually close to the sensory organs for senses such as vision. The brain is the most complex organ in a vertebrate's body. In a human, the cerebral cortex contains approximately 15–33 billion neurons, each connected by synapses to several thousand other neu-

> rons. *These neurons communicate with one another by means of long protoplasmic fibers called axons, which carry trains of signal pulses called action potentials to distant parts of the brain or body targeting specific recipient cells.* (Wikipedia)

The following pages show pictures of the human brain. This structure, located inside the skulls of *Homo sapiens*, is the most advanced and highly organized structure of the evolutionary process. At birth, the brain is capable of many highly complex functions. These include control and coordination of motion and balance, heart and respiratory functions, and the gastrointestinal system. The infant brain plays a part in the homeostatic mechanisms that keep the entire body functioning on a smooth and predictable level, plus a myriad of other simple and complex things. As the body continues to grow, the brain continues to grow and develop, receiving a constant flow of data from its senses.

The cerebral cortices, by far and away the most complex part of the brain, are responsible for "thinking"—that is, consciousness, recalling data stored in memory, originality, emotions, analyzing, processing information received through the senses, plus many things too numerous to list. In a real sense, we "are" our brains.

The essay following these pictures is an attempt to show that the brain, in spite of its incredible abilities, is limited and not capable of understanding everything about the natural world in which it finds itself. And in addition to not being able to understand fully is the problem of putting into words or images the real picture of what enters our consciousness through our senses. This makes it extremely difficult to remember, replicate, and transfer this information to another human brain. And realizing, of course, that our brains are the structures thinking that our brains are incredible, which opens up a new area for consideration.

All that we humans "know" about the universe in which we live is simply what our brains understand. Our brains, over a lifetime, accept the impulses (that is, light, sound, odors, touch, taste, etc.) from the world around us and make interpretations and react to them. Most of this comes through our visual and auditory senses by reading or hearing the accumulation of knowledge passed down by other brains.

THE HUMAN BRAIN

We thereby increase our understanding of our universe. This in no way, using this limited apparatus, means we know all the realities of the world in which we find ourselves. We (our brains) simply don't have the capacity. The development of computers and other instruments that help us gain data certainly increases our understanding, but always, in the final analysis, the human brain is the final arbiter. If the human brain succeeded in developing an "artificial intelligence" that surpassed the ability of its creator, the human brain would still be limited. There is always the theoretical possibility that human brains could develop methods of attaching electronic or other devices to their brains that would increase the brain's ability to interpret the universe. But that's another story and will not be considered in this presentation.

The following is a picture of the human brain. This is where we reside. All the anatomy, physiology, chemistry, nutrients, and every other function of the entire rest of the body serve only one purpose: to supply and protect the brain and aid in its function.

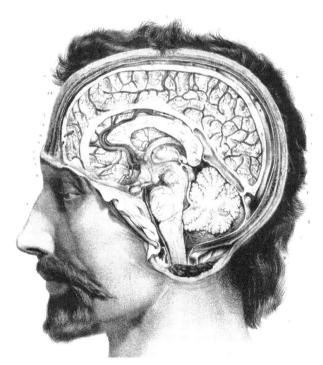

Frank Netter's CIBA Collection of Medical Illustrations.

The following schematic shows how the human brain receives new data or information and the general area of the brain where the information is received and stored and analyzed. There are multiple theories and claims that the brain receives new data from sources outside of these presented and outside of the consciousness of the individual.

Input into the Brain through the Five Senses

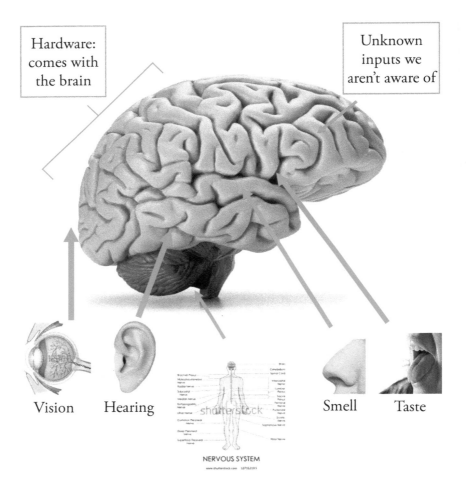

(Peripheral nerves carrying inputs from the rest of the body to the spinal cord and then to the brain.)

THE HUMAN BRAIN

Whatever our opinions may be about evolution, we all can agree that the end product of evolutionary process regarding human beings ends with the brain. All the other body parts—the musculoskeletal system, the gastrointestinal tract, the cardiovascular and respiratory systems, liver, pancreas, etc.—are merely systems to support brain function. No brain, no person. But at the same time, the brain is dependent on the body for oxygen and nutrients, as well as the sensory organs, to allow it to function. The brain and the body are interdependent in their physical aspects, controlled by the natural laws. The two make up the individual.

The brain and the body have locomotion, which gives the combination the ability to perform physical acts, including exposing the brain to continuous sensory inputs. This gives the brain a freedom which it would not have otherwise. An isolated brain preserved and kept alive with its sensory inputs by artificial methods would be totally dependent on its keepers to supply the sensory inputs and, therefore, deprived of any freedom to make decisions on its own, thereby limiting its development.

And of course, the health and well-being of the body affect the function of the brain. Lung and heart disease lower oxygen supply to the brain. Gastrointestinal maladies affect the nutrients delivered to the brain. Liver and kidney disorders allow an accumulation of toxins that affect the brain. Vascular abnormalities can limit brain function. And always at work, our old nemesis, the second law of thermodynamics, dictating gradual disorganization.

This chapter describes the structure and function of the human brain and its capabilities. Importantly, we will also look at its limitations—it's not that there are some things that our brains "don't get"; it's that there are some things that our brains aren't capable of "getting." The function of the human brain is limited and is incapable of understanding everything that is presented to it through its senses or is brought into its consciousness.

A case can be made that the human brain is capable of understanding much more than the brain can bring into consciousness. In certain circumstances, such as dreams and the phenomena described by those who have had "near-death experiences," these ideas and "facts" enter the consciousness. It appears, especially in the case of dreams, that our brains contain ideas that come into our consciousness during sleep and the period shortly after that that are contained in the physical makeup of the brain, and for unknown reasons spring into consciousness. The ideas and beliefs about the origins and meaning of dreams are multitudinous and go all the way back in history to the time of the ancient Greeks and Israelites. Maybe dreams are simply the brain cleansing itself of all the debris and unnecessary electronic signals and impulses and chemical waste developed during the waking hours. While the neurons are being rearranged and put back in order, random, meaningless thoughts enter our consciousness.

The following thoughts and ideas are presented to further emphasize how complex the human brain is, and to support the position of how improbable it is that the human brain developed through random processes.

The human brain is constructed in such a way as to think and comprehend in terms of length, width, height, and time. Try as we may, we cannot think of anything that does not have these dimensions. We can be aware of transcendent things—that is, the things that aren't physical: emotions, love, hate, joy, humor, etc. We cannot fully comprehend or understand them, but we surely know they exist. But when we analyze our physical world, we are limited by (imprisoned by) our inability to think in terms beyond the four dimensions mentioned above. These dimensions are a function of the human brain. Humans cannot comprehend an infinite period of time or an infinite quantity of space. It's not that we aren't smart enough, but that the structure that we are using, the human brain, is not capable of understanding these things, just like computers have limitations. It may be there are other brains out there whose physical makeup is entirely different and have no problem thinking in terms of what we humans call infinite time and space. *The problem is not the complexity of the physical world but the limitations of the human brain.* Sooner or later the brain "hits the wall" and can go no further.

And look how difficult it is to explain consciousness. When the human brain consciously tries to understand consciousness, it comes to a standstill: hits the wall, can't go any further, brain limit. See the chapter on "Consciousness." You will note it is a short chapter and not very enlightening because the human brain has yet to grasp even the basics of understanding consciousness.

As further support for the concept of brain limit, note that some of the greatest scientists and thinkers throughout human history were convinced that the human brain is limited and incapable of understanding all reality. The following sentence is abstracted from the quote on the mystical by Albert Einstein in the introductory pages of this essay. As he said, *"To know <u>what is impenetrable to us really exists</u>, manifests itself as the highest wisdom and the most radiant beauty which our dull faculties can comprehend."* The point of this chapter is to look at the human brain from the point of view of its limitations, to realize as Dr. Einstein did that *"what is impenetrable to us really exists,"* that our brains are incapable of understanding all reality.

And again Dr. Einstein…

> *The problem involved is too vast for our limited minds. We are in the position of a little child entering a huge library filled with books in many languages. The child knows someone must have written those books. It does not know how. It does not understand the languages in which they are written. The child dimly suspects a mysterious order in the arrangement of the books but doesn't know what it is. That, it seems to me, is the attitude of even the most intelligent human being toward God. We see the universe marvelously arranged and obeying certain laws but only dimly understand these laws.*

Michael Polanyi (a Hungarian-British polymath, who made important theoretical contributions to physical chemistry, economics, and philosophy, and a significant influence on John Polkinghorne, the Cambridge physicist-turned-Anglican priest):

> *It is not enough to believe that ideals such as truth, justice, and beauty, are objective, we also have to*

> *accept that they transcend our ability to wholly capture them. The objectivity of values must be combined with acceptance that all knowing is fallible.*

For the human brain to understand large phenomena, it must break them down into smaller pieces that it can understand, analyze, and reassemble (synthesize). When we have broken it down as far as we can go, we have reached brain limit. This doesn't mean that it cannot be broken down further, but that we don't have the ability to do so.

Our brains exist in a space that is filled with of all sorts of waves: light waves, sound waves, ultraviolet waves, radio waves, TV waves, telephone waves, and numerous others that we are not aware of, as well as many that we are aware of and use. With the right instrument, we can receive many of these waves and convert them into images that our eyes and ears can receive and transfer to the brain. Are there other "waves" or impulses that enter the brain through processes that become part of our consciousness? Do some brains pick up signals that other brains do not, or that some brains are incapable of picking up? Do these signals come from "on high" as the spiritually minded would say, or are they part of the natural world that we simply don't comprehend? Does our ability to input these signals increase with knowledge or experience? Does our ability increase with a "developed conscience"?

Is there an Outside Force influencing our brains? A force operating outside the natural laws as we understand them? Influences that we are not conscious of?

Charles Darwin himself had reservations about our ability to understand. The following is a quote from 1860:

> *I am inclined to look at everything as resulting from designed laws, with the details, whether good or bad, left to the working out of what we may call chance. Not that this notion at all satisfies me. I feel most deeply that the whole subject is too profound for the human intellect.*

Using our brains to think about our brains has obvious limitations from the very beginning. Is it possible that any entity built of the physical elements of the universe, and based on reason and analysis, can completely understand itself? Just considering animal brains is beyond us. There are thousands of examples. Some examples are the following:

Ants on land can sense a bread crumb on a boat and traverse a fifty-foot dock and climb down the ropes to the boat for the food. This is done with a brain the size of a grain of sand. Detecting food at such a distance and navigating their way down a dock and a rope for this food is beyond our brain's ability to comprehend. Compared to the most capable of human electronic computers relative to size and ability, the ant's brain is incomprehensible to us.

Birds with brains ranging from the size of BB shots to walnuts can easily navigate long distances, thousands of miles, to return to their previous nests and then back home again, with the built-in capabilities of their brains. This is beyond our brain's ability to comprehend.

And of course, the function of the brain depends on structure and anatomy (just like with mechanical things—an engine, for example—depends on the structure). There can be no change in function without a change in anatomy. They are inseparable. Function cannot change without a change in the structure. The evolution of the human brain, just as with the rest of the body, is dependent upon a change in the structure. The evolution of the human brain, developing billions of cells, synapses, coordination with the spinal cord and the senses and the rest of the body, with no guiding principles or forces, for no reason except by chance, stretches our understanding of probabilities and randomness.

The questions and challenges related to the generally accepted theory of evolution of the human body, by chance mutations, are multiplied many times over by the development of the modern human brain. Did the blood circulation, oxygenation, gastrointestinal tract, providing nutrients develop first and then a brain developed? Or did the brain cells develop first and needed circulation which then developed in response to this need? Darwin would say that mutations lead to changes in structure. Determinists would say it was determined at the moment of the big bang. The religious would say it was caused by an Outside Force. It is essential that we develop an explanation for the status of the human brain, as it exists today, that is far more reasonable and acceptable than the accepted Darwinian theory.

The rest of this chapter is devoted to listing and discussing some of the other functions and abilities of the human brain, again to add additional insight into the overall complexity of the brain. These are brain functions that are not well understood and are hard to correlate with the physical structure of the brain as we understand it.

The phenomenon of intuition, although not well understood, is accepted by scientists, as well as laypeople. By definition, *intuition* is (*Dictionary.com*):

> *A phenomenon of the mind, describes the ability to acquire knowledge without inference or the use of reason. The word intuition comes from Latin verb intueri translated as consider or from late middle English word intuit to contemplate. Intuition is often interpreted with varied meaning from intuition being glimpses of greater knowledge to only a function of mind; however, processes by which and why they happen typically remain mostly unknown to the thinker, as opposed to the view of rational thinking. Intuitive decision making is far more than using common sense because it involves additional sensors to perceive and get aware of the information from outside. Sometimes it is referred to as gut feeling, sixth sense, inner sense, instinct, inner voice, spiritual guide, etc.*

The possibility of there being a phenomenon as described above raises multiple interesting thoughts and theories. The main interpretation for purposes of this discussion is this: the brain can bring into consciousness ideas and thoughts that are not specifically a result of its physical structure plus data introduced into the brain through the senses—that is, the brain is capable of creating ideas and thoughts that exceeds "the sum of its parts." The brains of some individuals seem to contain "sensors" that operate on a subconscious level, receive and analyze data, and then transfer this data to the consciousness of the individual. This is basically what we traditionally call inspiration: "*a divine influence or action on a person believed to qualify him or her to receive and communicate sacred revelation*" (*Merriam-Webster*). Or are these sensors or senses simply a combination of all the other abilities developed or information gathered over time?

No discussion of brain function would be complete without an analysis of the common and somewhat overused word—*psychological*. When used properly, the word *psychological* simply refers to "something occurring in the mind." By definition, the mind is the element or complex of elements in an individual that feels, perceives, thinks, wills, and especially reasons. However defined, the "mind" is still part of the brain, functioning and operating according to the same processes as the rest of the brain. If a thought and subsequent action is created by stress or tension, hypnosis or suggestion, it still operates within the confines of the physical brain. Frequently, the word *psychological* is used as if the mind was operating outside of the physical structure of the brain, producing thoughts and actions as if they came out of thin air. This is acceptable, even comforting, in casual conversation, but for purposes of this discussion of the brain, its abilities and limitations, we will assume all brain functions are dependent on the physical makeup of the brain. There is no such thing as magic. There is always an explanation. Nothing happens without a reason. So with brain function. (At this point we do not discuss the idea that an Outside Force or some other influence, outside of the senses, may enter the brain structure and create ideas and changes. This issue will be addressed in chapter 16, "Summary.")

I relate a personal experience, one shared by many laymen in folklore history, but persisting and defying objective analysis. Over the years I have had many direct experiences with the phenomenon of "talking off warts." My nephew had a small wart on his lower eyelid. This growth was treated by freezing and cauterizing with negative results. It was then meticulously removed by plastic surgery and promptly recurred. The local barber in my hometown had a reputation for talking off warts. He placed his hand over my nephew's eyelid and mumbled a few unintelligible words. Within a week, the wart disappeared and has never recurred. Another member of my family had several warts on her fingers. Over a period of years these warts were treated by freezing, chemical destruction and surgery, each time with recurrence and persistence of the warts. The hometown barber performed his magic, and within a week or so, all the warts disappeared and never returned.

Intrigued by this, I did my own experiment. I "treated" about eight to ten "patients," adults and children, who had warts on various parts of their bodies. I used a different method with each person. Sometimes I covered the wart with a surgical bandage in a dimly lit room and mumbled some incoherent phrases. I sometimes stared straight ahead and pressed on the wart, and one time I just walked into the treatment room and told the person that the warts would go away in a few days. In four or five cases, the warts disappeared within one to two weeks. Unfortunately, I did not keep any records or follow any strict protocol. But I am convinced, by the experience of my family members and my short study, that there is a real possibility that "suggestion" is a real phenomenon.

The modern textbooks of dermatology list suggestion as a treatment for warts. My theory as to why this technique works is this: warts are caused by viruses. The actions of the person making the suggestion enter the person's brain through his visual and auditory senses. These impulses are then sent to that part of the brain that governs the immune system. The immune system responds by creating antibodies against the wart virus. The antibodies destroy the virus, and the wart is cured. This explanation is compatible with my basic point that there is always an anatomical cause for brain function unless acted upon by some Outside Force. In the case of warts disappearing, the brains of the patients that responded did so because these brains were anatomically and neurophysiologically constructed to respond to the audio and visual inputs that caused the reaction. Why some brains are capable of this response and others aren't is not for this discussion. But this phenomenon is in no way related to the "Outside Force" as used in this treatise.

And of course, the age-old and poorly understood phenomenon of the action of placebos. A placebo is a substance or treatment with no active therapeutic elements. Common placebos include inert tablets (like sugar pills), sham surgery, and other procedures that clearly should have no obvious benefit. However, so many patients respond positively to placebos that the name *placebo effect* is included in the medical literature and placebos are required as part of any legitimate investigation of new drugs or procedures. They work so often that many, if not a majority, of otherwise objective people believe their effect must be accepted as legitimate. Placebos are a poorly understood phenomenon, but cannot be dismissed.

The phenomena of the psychological "talking off warts" and the placebo effect are included to show how complex the human brain is and add to the case that chance mutations and random evolution would never produce such a complex entity.

Another thing to keep in mind (or in your brain) is that the brain, from the day it comes into this world until the day it dies, is constantly flooded with new inputs from the senses. The infant hears, sees, and feels things from the moment he first awakes on birth. This information is stored in the brain's memory banks. Much of this will prove irrelevant and useless. Much of it will be forgotten, erased. Much of this may appear to be irrelevant, but it affects the brain's understanding of any given topic, although not specifically related to it. But this constant buildup of data, experience, repetition, and continuous "on-the-job training" results in a brain that constantly increases in ability and competence.

And how about the use of symbols, pictures, and abbreviations? Every book, magazine, article, or even this book you're reading now is composed of nothing but the twenty-six letters of the alphabet plus a collection of symbols and spaces, all arranged in a different order. The human brain subconsciously in a flash recognizes and gives meaning to the millions of images presented to it through its visual senses. These are immediately brought into consciousness by the brain, with many of them remaining subconscious, but still having an effect on understanding and behavior. You may not understand or be conscious of the meaning of everything you read or see in a movie or a picture, but a significant amount of all this information enters the brain and affects its function and understanding on a subconscious level. It's possible that what you are now reading in this book influences you in ways in which are not aware.

And how about the fascinating phenomenon of savants (sometimes called idiot savants)? A *savant* is defined by *Merriam-Webster* as *"a person affected with a developmental disorder (such as autism or mental retardation) who exhibits exceptional skill or brilliance in some limited field (such as mathematics or music)."* Darwinian theory could explain this by postulating that the savant evolved enough to survive in a closed society, surviving with the help of those more advanced. This, of course, is true of all humans who have limited capabilities. The brains of savants differ in that their brains have specific skills that exceed the skills of "normal" brains while lacking in the usual abilities that we define as "normal."

A somewhat similar phenomenon is that of the otherwise "normal" human brain which is capable of performing tasks or feats spontaneously, without any prior effort or practice. Most of us are familiar with the person who can play the piano "by ear" without any training or practice. The person's brain appears to have been born with the ability—the hardware came with the brain, whereas other brains have to "load" the software, learn and practice.

There are many reported cases where brains after significant injury perform tasks that, before the injury, they never expressed, and surprised even themselves with their abilities. And sometimes they are capable of feats normal brains cannot do. In most cases, this phenomenon occurs in brains whose injuries are serious, with neurological loss, even debilitating. Some researchers believe that the sudden expressed ability was already in the makeup of the particular brain and that the injury removed an inhibition and allowed the ability to express itself.

These examples add to the position as described in this chapter—that is, the human brain is an extremely complex organ, made up of the elements of the planet, with many abilities, but also by its very nature, is limited. These examples also support the case that efforts by humans to create a similar structure—that is, artificial intelligence, are pale and rudimentary versus the complexity and ability of the human brain. We may develop robots and machines that confront us or even destroy us. But that would be a great failure of the machines—doing the opposite of what they were created for.

And a sci-fi closing thought on this chapter: Perhaps there are other signals, impulses, or wave frequencies throughout the universe, including the planet Earth, that can be and are received by the human brain and converted into conscious thoughts or subconscious actions, even though we are not aware of receiving these impulses. These would become a part of each individual's brain just as are the continuous inputs from the senses. Could these extraterrestrial impulses account for intuitions, the creations of the great works of art and music that seem to exceed human understanding, the unique abilities of certain brains throughout history to originate ideas and thoughts that seem to be beyond the normal human brain function? Why do some individual brains respond to these impulses while most do not?

Could these signals create our awareness of transcendent things? This, of course, creates horrific problems, such as why do some brains receive the signals and some do not. This suggests that some humans would be incapable of responding to the impulses that create transcendent feelings. This idea would still leave room for the concept of free will, as an individual who receives the signals would still have a choice to accept and act on them or not. However, it would suggest that we have two distinct types of human beings—those who are capable of receiving transcendent impulses and those who are not.

Human beings cannot detect these signals with any of our technologies and are therefore inclined to deny their existence. But we are forever detecting and creating new types of waves in our physical world, so is it far-fetched to theorize the existence of those impulses described above?

Chapter Five
Science and Darwin's Theory

The most useful piece of learning for the uses of life is to unlearn what is untrue.
—Antisthenes

Condensed: This chapter provides a generalized description of the methods of science and its relation to development and understanding of Darwin's theory of evolution. It points out the limitations of science in general and how new information has changed many accepted scientific principles that have endured for centuries. Modern science realizes these limitations in all areas of study, including Darwin's theory. That is why Darwin's theory is still called a theory.

*****

*N*o study of Darwin's theory of evolution would be complete without discussing science in general. Also, any study of evolution and Darwin's theory must address how it affects so much of our thinking in areas other than pure science: philosophy, religion, politics, business, and economics. A part of this presentation will be an analysis of Darwin's theory of evolution, and specifically what we should teach our youth. Included is a discussion of the limits of science, and specifically the weaknesses of what is generally presented as human evolution—that is, Darwin's theory with its emphasis on evolution by chance mutations and survival of the fittest. The method

SCIENCE AND DARWIN'S THEORY

used will be to present a variety of topics somewhat haphazardly with an analysis and opinion, and then conclude with a synthesis.

The essay will refer to two major areas of thought—science (objective) and transcendent (nonobjective and not provable). The word *transcendent* is used to mean all areas of thought that do not lend themselves to the scientific method of investigation—that is, those things that cannot be proved by the usual scientific methods.

In studying the development of the universe and the planet Earth's place in it, we have four choices:

1. Creationism—meaning the creation of the entire universe by a God or some divine power in the relatively recent past, including the creation of planet Earth and human beings in their present form.
2. The big bang with all that followed determined (determinism) according to a system of natural laws.
3. The development of human beings and animals on the planet Earth according to natural laws with modifications by chance mutations (Darwin's theory).
4. Evolution affected by an Outside Force, or forces, which we cannot fully comprehend or prove their existence.

Why the endless arguments and conflicting views regarding Darwin's theory? *Because it is about the very basis of how one views life itself. It is about much more than science. It truly is where science and religion and philosophy intersect.*

The arguments—pros and cons—over evolution and Darwin's theory have existed for over a century and a half. It is safe to say that most scientists agree with Charles Darwin's theory, but a significant number of learned people disagree. Like many other conflicting opinions, there appears to be much truth on each side of the argument. Like the old professor who was asked in the midst of a heated discussion among several professors, what his opinion was. His answer was, "I agree with everything said so far." And then he explained that when an argument goes on for a long time with conflicting views by well-informed people acting in good faith, then there is almost always an element of truth on all sides. Our task as we progress is to develop a synthesis of all opinions that have elements of the truth.

To begin with, let us try to establish what part the human brain plays. But can we do that? There seems to be no getting around the point that everything we try to do intellectually will always go back to the point that we are using our brains to study the object—the brain. This has always been a problem for humans since the beginning of philosophical thought: we are using our brains to try to understand our brains.

Is it that we don't have an answer, or that we cannot find an answer? Or that we are not capable of finding an answer? That our brains will always at some point "hit the wall"? A basic point of this treatise is exactly that (see the chapter on "Brain"). But we are capable of a better analysis and can come closer to the truth regarding the subject of human evolution.

The primary goal is to help laypeople realize the limitations of science, and that although there are no presently outstanding scientific alternatives to Darwin's theory, there is certainly room for vigorous discussion of the problems of Darwin's theory. The science and the physical evidence is there to support the theory of the evolution of human and animal bodies in some way. The evolutionary processes may not be exactly as articulated by the modern scientific and historical communities, but there is a great deal of evidence to support what is commonly called evolution. The challenge is to Darwin's theory as to cause: that the evolutionary changes in human beings and animals are the result of random mutations and survival of the fittest. And the almost universal acceptance of, and teaching of, evolution synonymously with Darwin's theory.

The challenge is to make the case that chance does not play the determining role in evolution, as the word *chance* is commonly used by scientists and laymen. That analysis using chance and randomness is limited and always leaves room for error. This basically could be viewed as a third theory to explain evolution, along with Darwin's theory and Creationism. Or otherwise stated, there are three reasons for evolution: (1) Darwin's theory, (2) Creationism, and (3) some Outside Force that we cannot measure or fully understand.

Let's begin with this basic point—that everything is always changing. Our measurements of time, the continuous motion of the bodies in the universe, the continuous decay of the universe, our consciousness, etc. Why? There's a reason—there's always a reason for everything; nothing happens for no reason. There is always a force or impetus to cause change.

So what is the force, or forces, that cause change? Why does change occur? Isaac Newton's first law of motion states, *"When viewed in an inertial reference frame, an object either remains at rest or continues to move at a constant velocity unless acted upon by an external force."* Newton's law describes physical objects and their behavior. Something has to cause a change. Does this law or a similar set of laws bear on every type of change in the universe? Is it a single force, or is it multiple forces? Can we understand? Is the human brain capable of understanding?

And what part does free will play in all this? Determinism denies the possibility of free will. If one entity acts with free will, it will affect other entities and cause a change in course. The old philosophical argument—if A had not decided to drive fast, B would not have been killed. If B not been killed, then C would not have married D and their offspring would be different, etc.

Another point to consider is this: Who is to have the final say in conflicting and contradicting subjects? In Prof. Lawrence M. Principe's *Great Courses*: *Science and Religion*, he emphasizes the issue of authority. For example, who speaks for science? Who speaks for theology? How does one decide between rival scientific interpretations or between rival theological perspectives? How is authority gained, and how is it respected, or not respected, in an ideal world and in the real world? How does one achieve the ability to speak authoritatively about scientific issues? About theological issues? Do the answers to the previous two questions differ? Should they?

With the massive amounts of information available today through literally hundreds of sources, with endless forums available to allow almost anyone to give an opinion on any topic, we must commit ourselves to truth and accuracy. We must not become cynical, but always be aware of the possibility that something that is published, printed, televised, or appears somewhere on the Internet may not necessarily be accurate. We must be constantly on guard against the possibility of being manipulated intentionally. One of the biggest problems facing modern human beings is to find a way to filter out the truth from these multiple sources.

Developing an understanding of a subject, whatever it might be, almost always has to be done in steps. You must understand *this* to understand *that*. To understand how a computer works, you must have knowledge of electricity, electronics, algorithms, logic, engineering, etc. In other words, you use an evolutionary process from one step to the next. It appears to me that this is what we do when we study the basic physical evolutionary changes of man and animals. There are clearly evolutionary processes. However, Darwin's theory goes beyond the basics and tries to answer the "why."

Chapter Six

General Thoughts
Science, Teaching, and the Transcendent

Human beings live in a realm untouched by science. It is the domain of appetite, passions, sentiments, aestheticism and morals. The only truth in this realm is what we perceive and what we feel.
—Pierre Lecompte du Noüy

Condensed: This chapter looks at the problem of passing on information from one generation to the next—that is, the teaching of students. There is a strong consensus as to the teaching of scientific subjects—that is, the teaching of things that can be proved using the scientific method with its emphasis on reason, analysis and proof. Almost everyone agrees that almost every human being has beliefs and feelings about many things that cannot be proved. The teaching of the arts and other aesthetic enjoyments is universally accepted as appropriate for public education. The teaching of the idea that there is something that is beyond human abilities to fully comprehend or analyze, the transcendent as it were, in public institutions is considered dangerous because it is viewed as the teaching of religious faith. We all agree that the teaching of religion is not a public issue. In this chapter, I try to point out that no public education is complete without making sure that the students understand that almost all human beings, including scientists, believe there is something beyond those things that we can prove in the laboratory.

*****

As we grow older and our knowledge and experiences increase, our understanding of the world in which we live evolves. In my youth and early adulthood, I believed that science, religion, philosophy, art, history, entertainment, and sociology were all basically subjects unto themselves, with some overlap, but remaining distinct. Each was represented by its own circle. Then I began to believe that these circles start to overlap with increased knowledge, experience, and wisdom. Each of these circles became bigger, but at the same time the overlap of each became greater. I postulated that the circles would finally totally overlap each other, as each entity understood the other entities more completely. I now think that a better understanding is to view the advancement as a series of liquids being poured into one container and becoming a solution, each entity becoming a part of the whole, each influencing all the others in complex and unknown ways. Not like a candy bar of mixed nuts, sugars, syrups, but a solution of liquids. As our brains develop and gather more and more information and experiences, we are able to understand things without being aware of the processes going on in our brains—subliminal (existing or functioning below the threshold of consciousness).

One of the essential goals of education is to teach the truth, and to declare that transcendent things are not true and not worthy of study is a failure of the point of education. The challenge is to teach the possibility of transcendent things without teaching religion or proselyting because this would violate the Constitution and the laws of the United States, as well as opening a door to multiple conflicts about definitions and beliefs.

It can be done with the following understanding and restrictions. We all agree with the present methods of teaching science with a strict adherence to objective and rational thinking and accepting those things that are proven true or are in the process of being proven. The objection is that we do not teach the younger generations the *possibility*

GENERAL THOUGHTS: SCIENCE, TEACHING, AND THE TRANSCENDENT

of transcendent things. That doesn't mean teaching those things that are generally accepted as being disproved by modern science.

What we can do is make sure that the younger generations, as well as all of us, understand that there are many things that cannot be proved, using modern science, but are *possibly* true. These are the transcendent things: things that have stood the test of time and rigid analysis and objective and analytical thought by the brightest of scientists and other scholars over the centuries and by their very nature cannot be proven, but neither have they been disproven. This certainly means that there is a real possibility that they are true, and as we commented earlier, an essential ingredient of education is to teach the truth.

Certainly another essential ingredient of education is to help human beings live more enjoyable and fulfilling lives. Even the strictest scientists, most rational thinkers, and the most unemotional among us believe being happy and enjoying life is a positive good. No one would deny that it is proper to teach art, literature, aesthetics, and the social sciences. These are beyond our ability to prove their value. They are transcendent. But we certainly believe that they are highly desirable and, though not necessary for survival, should be taught. History, math, and the hard sciences are taught with everyone agreeing they are necessary for pragmatic reasons.

We can argue forever about what should be required in our public schools and universities. It is distressing to see the marked change in what we now see as required and necessary parts of an education. But certainly it's not too much to ask that when teaching any subject that is a theory, even though generally accepted, that the student be made aware of the controversies and shortcomings and emphasize the importance of continued study and keeping an open mind. I think this is extremely important in the teaching of the *theory* of evolution.

You have to address the issue of transcendent things—they do exist. You cannot deny faith as nonexistent. In keeping with my basic point that nothing happens without a reason, and in the case of transcendent things, there must be a basic anatomical and neurological reason or an Outside Force influencing the brain. To claim they exist only in the human brain and do not exist outside of the human brain needs an explanation as to why they developed in the human brain to begin with, and what part they play.

It is also true that many people who believe that they respond only to the rational and objective have strong feelings about their beliefs and positions. Atheists and secular humanists are often outspoken defenders of their beliefs. They have seminars, journals, and use the media, social and otherwise, to promote their beliefs. Why promote and advocate atheism? Because they have beliefs. They have gone beyond analysis and conclusions and now feel strongly that they must convince others of their beliefs. Proselyte, as it were. Why is that? There is no scientific reason. It must be something beyond the usual difference of opinions. We all are influenced by ego and pride, but I think these impulses go much deeper than that. Why the universal, almost insatiable need, to convince others of our beliefs?

It is interesting, and also terrifying, to consider what part education and experience have to play in human's beliefs about transcendent things. What can we say about scientists and philosophers who are more learned—that is, have accumulated more data, trained in analysis and more generally competent in intellectual matters? Are they better able to understand theology or First Causes than the average person? It is comforting and reassuring to note how most, the learned and the unlearned, come to believe the same in these important areas. Others become extreme fundamentalists and some become atheists. It is intriguing, yet disturbing, to see how someone like Stephen Hawkins becomes more and more convinced that there is no God or outside influence in the natural world, the more he studies and gathers data. On the other side, the history of mankind is filled with highly intelligent people who do believe in transcendent possibilities.

All of us as individuals must think and study and contemplate and develop our own beliefs and not carte blanche accept or reject any ideas regarding transcendental things because someone who is "intelligent" or "scholarly" holds those views. The logical end of such a position would be, "If I were smart enough, I would be an atheist or a believer."

Chapter Seven
Chance and Probability

> *Our scientific laws are statistical laws; we explain them by the calculus of probability... These statistical laws enable us to foresee phenomena even though we do not understand how they work or their cause... We have to limit cause to the preceding event. However, when man intervenes, it is generally simpler to consider his thought as the efficient cause. A personality appears able to influence events, but these statistical laws cannot take personality into account and a hypothesis is needed to bridge this gap.*
>
> —Lecompte du Noüy

Condensed: This chapter is perhaps the most important chapter in this book. The use of probabilities is a very important part of general science, experimentation, and analysis. Darwin's theory of evolution is completely dependent on changes due to "chance mutations." Any understanding of Darwin's theory must understand the use of chance and probabilities and cause and effect. This chapter presents a very general description of the very complex subject of chance and probability. My biggest challenge to the generally accepted theory of human evolution is to the way chance mutations are understood in the overall scheme of how chance and probability are used in general science.

*****

> Chance: *the absence of any cause of events that can be predicted, understood, or controlled: often personified or treated as a positive agency: Chance governs all.* (*Merriam-Webster*) (Author's note: this definition says the absence of any cause that can be "predicted, understood, or controlled"; it does not say that there is no cause.)

*I*n science, as well as almost everything, humans can never completely prove everything and therefore accept probability (and its cohort, chance) as the best we can do, knowing that there's always the possibility of error. The biggest problem in the search for the absolute truth about our origins is how far back we go. With our present brains and research capabilities, we can only go back so far and almost always one step at a time. We see that one discovery has led to another to another to another, as we move progressively backward in time. The scientist is satisfied with ending his investigation at the time of the big bang. In trying to establish the absolute cause, we have to go back before the big bang.

The word *chance* is used in a variety of ways, with different meanings to different people under different circumstances. Many disagreements are purely semantical. It is important to understand that when scientists use the word *chance*, they basically mean that they are satisfied with researching a particular idea back to a point where it pragmatically supports their thesis. They do not need to go further back because they have all the material they need to proceed confidently that from now on, their theory will prove to be reliable and predictable. They are content to say that probability brought the process to this point.

It is very important to remember that this is the position of Charles Darwin. Darwin believed that a "chance" mutation produces a new species that can survive in a new environment, as the prior species dies out. He proved to his satisfaction that this was true by the artifacts he studied. From the study of this singular phenomenon, he proceeded without considering the processes that caused the mutation and delegated it to "chance," as a "chance mutation." *We must remember that Darwin's theory of evolution did not make any claims about a primary cause.*

The scientist has no need to go further back or be concerned with a primary cause, or a complete understanding of the cause or causes once his theory proves workable, reliable, and reproducible. But our age-old impulse, even obsession, with finding the ultimate truth and meaning compels us to go beyond the point that satisfies the objective goals of the scientists. We are filled with a tension that will not let us rest until we find an answer to the question of the ultimate cause or reason for our existence. That we are here due to the processes of "chance," with no need for an explanation of a primary cause, leaves us cold and dissatisfied. The rest of this chapter is my attempt to convince the reader that there is a cause for everything. That means, of course, that there is a cause for our existence. And the driving force behind all philosophies and religions is to find and understand that cause.

As I say throughout this treatise, nothing happens without a cause. If taken back far enough, the cause of any action can be explained. One of my favorite, somewhat humorous, examples is the phrase "the gun went off." This phrase is used in many news articles and police reports, stating things, like, "He was cleaning his gun, and the gun went off." "He thought the gun was not loaded when it suddenly went off." "The gun was lying on the table, and it went off." It appears they are suggesting that was no cause for the gun to fire. It just went off.

As Lecompte du Noüy described in his book *The Road to Reason*,

> *Every event has a cause and, more often, several causes. In the case of a cannon shot, for instance, shall we say that the firing of the shell is caused by the explosion of the percussion cap, or by the movement of the hand that pulled the string? Shall we say that the cause is the charge of powder? But without the movement of the hand, the charge could have remained inert for centuries. Nowadays all mechanical movements can be amplified and we may eas-*

ily imagine that the explosion of the percussion cap, obtained by electrical means, was originally brought about by a feeble ray of light that could have been intercepted by the wing of a fly. We could have projected the ton of steel, which the shell weighs, thirty miles away by harnessing and amplifying a ray emitted by Sirius. The Chicago exposition in 1933 was lighted by a feeble ray, only a few photons, emitted by Arcturus forty years before. In the case of the cannon shot it would seem absurd to make the star responsible for the havoc wrought by the shell, and yet this slender beam of photons will have played as important a part in the firing of the shot as the charge of powder. Neither can we say that the workmen who manufactured the powder, nor the chemical engineers, nor the owners of the factory, nor the inventor of the formula, nor his mother, father, etc., are responsible. And yet, all of them, all the men who contributed to the construction of the cannon, share the responsibility, which gradually crumbles away without ever disappearing completely and reaches back to the origin of the world. And I have not even mentioned the psychological causes, without which there would have been neither shell, cannon, charge, percussion cap, nor the continuity of coordinated efforts with the object of making the shot go off.

In studying causation, or seeking a reason why, we start with the final result and go backward to the point at which our need for proof, our curiosity, or our philosophy is satisfied. As stated elsewhere, the scientist is satisfied when the evidence supports his theory. The curious are satisfied with any number of answers and ideas and are comfortable frequently changing their position. Others must go on backward beyond the big bang, seeking basic causes. My contention is that there is no such thing as chance when taken back far enough. Nothing ever happens without a cause. All physicists accept

Newton's first law of motion: *"When viewed in an inertial reference frame, an object either remains at rest or continues to move at a constant velocity, unless acted upon by an external force"* (Newton's first law was referred to in chapter 3, "Science and Darwin's Theory of Evolution"). This, of course, pertains to the motion of objects. Can we assume that nothing in the universe changes unless acted upon an external force?

To summarize the above—if a process, even if it is disorderly or "random," if repeated over and over produces the same result often enough, then it can be considered predictable and, therefore, useful. In the case of flipping a coin, although the process is disorderly and unpredictable and different with each flip, the predictability of a fifty-fifty heads and tails split becomes greater as the number of times the process is repeated, as long as the variables remain constant. When viewed from this perspective and used properly and consistently and we all understand this method, then "there is such a thing as chance." But for purposes of this study, the point still holds that if you go back far enough, something has to cause change. An Outside Force? Again, nothing happens without a cause. But since in many areas of study we cannot go back far enough, we must use the methods available to us.

We must always keep in mind about how easily statistical methods can be misused, especially as to how they are presented to the general nonscientific public. So often analogies or connections or corollaries are made that seem reasonable, but on examination there actually is no cause and effect, producing confusion and contradictions. There is no greater area in the realm of human study in which there is such a spread in the extremes: the proper use of statistical studies and the improper use of statistics.

Let's look at the following illustrations to emphasize the difficulty in using statistics and evaluating random events and understanding the phenomenon we call "chance." I am not a statistician, and I am not challenging the value of the science of statistics and study of randomness. I am simply trying to impress upon the reader that the scientists and statisticians are interested in results that help them prove their theories. I am trying to show that statistics are not totally or completely "true" in the strictest sense, and therefore leave room for believing there is "something else" that affects changes and other phenomena, and a place for assuming there is an Outside Force affecting these entities.

Let's look at the example of flipping a coin, heads or tails. Flip the coin ten times, and you may get 80 percent heads and 20 percent tails. The more you flip the coin, the closer you get to a fifty-fifty split. Flip the coin a thousand times, and you expect a spread of approximately five hundred heads and five hundred tails. The reality is that a human being is flipping the coin. The fact that he places the coin heads or tails up on his thumb, along with many other factors (variables), plays a part in the final outcome. If the coin is heads-up on the landing surface, this plays a part in the coin's orientation when placed back on the thumb. How far away the coin lands, the number of steps he takes, the left foot or right foot first, fatigue, monotony, etc. play a part in the replacing and flipping of the coin. As this process continues from one flip to hundreds or thousands, the results of each flip remain inconsistent: it may land heads or tails. If the count is 500 heads and 499 tails, no sane person would bet his life that the next flip of the coin will be tails. But multiple experiments with human being's flipping a coin consistently produce an approximate fifty-fifty split after the number of tosses reaches a certain magnitude. It remains random and "chance," in the usual sense of the word.

Now let's create our own flipper machine—in a vacuum, a precise machine, a machine that reproduces the flip of the coin the same way each time onto a landing surface of a consistent texture. If you place a coin heads up on the flipper, and it lands heads up on the landing surface, then each time you put the coin heads up on the flipper, it will land heads up on the surface. Variables have been removed. The machine will continue to reproduce the same results until acted on by some Outside Force. It will not produce a different result because of "chance." If you believe that it does, then what is the force that produces the different result? (Friction, wear, and tear on the machine and the coin, construction flaws, etc. will eventually come into play, and our machine will become inconsistent. This is another topic and not part of this discussion. But this does not alter my proposal.)

The hard-to-understand nuance is this: In the first example of flipping the coin in the usual manner by hand, all the variables working with and against each other with enough flips (large enough numbers) will produce consistent results—the difference in the number of heads and tails is negligible, so that human beings can use these results successfully and without disaster except in the extreme of extremes. In the second example, that of creating a machine that consistently flips the coin, with removal of all the variables, we can predict from the outcome from the first flip. We have created an "Outside Force" that directs (affects) the outcome—that is, we have used human reason to direct and influence the outcome.

Also to consider—in the first instance, that of a human being flipping the coin, an individual human is involved, using his free will to flip the coin in any manner he chooses. But this holds true only with human beings exerting their ability to change things. In the natural world, including most scientific studies, human beings are not always involved, and therefore there is no element of possible change made by human beings using their free will. We can make the argument that humans are involved in all scientific studies and, therefore, introduce variables in setting up the physical parts of an experiment, as well as interpretation and analysis.

Let's look again at our flipping machine. Human beings are not involved in the mechanisms of flipping, although, of course, human beings built the machine. And this brings up another interesting question: The performance of the machine, the accuracy of the machine, and the longevity of the machine has been "determined" by its builders, in this case, human beings. Can we use this as a sort of analogy to the overall question of determinism? The performance of the machine was determined by human beings. But at the same time, the builders of the machine can "change" it in any way they desire.

Now let's look at throwing dice: If you use a machine that reproduces the exact same maneuvers for throwing the dice, you get the same result every time. Same goes for the roulette wheel.

THE DEATH OF ANNIE

A slot machine is somewhat different but the same principle applies. Whether mechanical or a modern electronic slot machine, the machine can be programmed to produce different results. It can be programmed to produce many winners or many losers and changes the so-called "chances." The casino operators make millions attracting customers by changing the machines to allow lots of winners to attract customers and then changing the machines to allow fewer winners.

Card games bring in a whole new element because two or more human brains are now involved, each making multiple choices that affect the outcome. However, the outcome is still determined by the choices made and is not dependent on some unknown Outside Force, or magic. The biggest winners in poker, playing many games over a long period of time as in professional poker, win not by chance but skill: they count the cards and remember which cards have already been played and, therefore, know that a certain card cannot be dealt because it has already been played, etc.

So I think we can make the case that there is no such thing as "chance" in the way that it is used in common conversation. Just as there is no such thing as magic. The magic trick always surprises and delights us, but there always is a clear explanation when we learn its secret. There is a reason for everything. When something changes or happens differently, there has to be a force of some sort acting on it.

All of a sudden something changes. Then this leads to more unexpected changes. This can lead to more and more chaos or to more and more order. It certainly appears that some things are determined (natural law) and some things aren't. Chance means that something happened outside of what has been determined or is outside of science. It happened for no reason. If it did not happen randomly, there has to be another answer. A force? Or determined from the beginning? Back to determinism. Or maybe at the beginning, it was determined that the free will of the human species was part of the process. But you cannot have a force (human beings) contradicting something which has been determined. Was chance determined?

The Darwinist example of the bird developing a bigger beak and then surviving in changing circumstances, while the small-beaked birds die out, makes a weak argument. The argument that chance causes a change in a chromosome that causes the development of a bigger beak does not speak to the question "Did a bigger beak not require multiple chromosomal changes?" A bigger beak requires more blood flow. which requires a bigger heart and vascular system and more oxygen. which requires bigger lungs and a larger GI system and bigger wings and more food to sustain the increased size, which defeats the purpose of the bigger beak, etc. And of course, my basic question—Why the original change? There is no change without a cause. Nothing happens without a reason. Does the Darwinist say change in diet, foreign substances in the water, a cosmic ray, electrons from outer space, etc. cause the multiple chromosome changes in all the organs? Is the answer "given enough time, anything can happen" science?

Another phenomenon that humans refer to is the thing called *luck*. It is frequently used interchangeably with *chance* by laypeople but is never used by the scientist. There is some legitimate overlap in the meaning of *luck* and *chance*, but for purposes of this discussion, let us be careful to keep them clearly separated. *Luck*, by definition, is *(1) the force that <u>seems</u> to operate for good or ill in a person's life, as in shaping circumstances, events, or opportunities; (2) good fortune, advantage or success, considered as the result of chance; (3) a combination of circumstances, events, etc., operating by chance to bring good or ill to a person.* From a strict, rational, scientific point of view, this definition and word is basically useless. But it is certainly helpful for most of us to be able to use the concept of "luck" to explain many of the happenings in our lives, as well as the lives of others. We, or they, were lucky or unlucky. To use this idea is frequently much more comfortable and reassuring than the truth.

Chapter Eight
Random

Coincidence is God's way of remaining anonymous.
—Albert Einstein

Condensed: This chapter follows up the previous chapter on chance and probability. It defines the concept of randomness and how it is used by scientists. One must understand the relationship between cause and effect to fully understand Darwin's theory of evolution.

*****

Random: *1. Proceeding, made, or occurring without definite aim, reason, or pattern: the random selection of numbers. 2. Statistics of or characterizing a process of selection in which each item of a set has an equal probability of being chosen. Randomness means lack of pattern or predictability in events. Randomness suggests a non-paid or non-coherence in a sequence of symbols or steps, such that there is no intelligible pattern or combination.* (*Merriam-Webster*)

Random events are individually unpredictable, but the frequency of different outcomes over a large number of events (or "trials") are frequently predictable. For example, when throwing two dice and counting the total, a sum of 7 will randomly occur twice as often as 4, but the outcome of any particular roll of the dice is unpredictable. This view, where randomness simply refers to situations where the certainty of the outcome is at issue, applies to concepts of chance, probability, and information entropy. In these situations, randomness implies a measure of uncertainty, and notions of haphazardness are irrelevant.

The fields of mathematics, probability, and statistics use formal definitions of randomness. In statistics, a random variable is an assignment of a numerical value to each possible outcome of an event space. This association facilitates the identification and the calculation of probabilities of the events. A random process is a sequence of random variables describing a process whose outcomes do not follow a deterministic pattern, but follow an evolution described by probability distributions. These and other constructs are extremely useful in probability theory.

Randomness is often used in statistics to signify well-defined statistical properties. Monte Carlo methods, which rely on random input, are important techniques in science, as, for instance, in computational science.

Random selection is a method of selecting items (often called units) from a population where the probability of choosing a specific item is the proportion of those items in the population. For example, if we have a bowl of 100 marbles with 10 red (and any red marble is indistinguishable from any other red marble) and 90 blue (and any blue marble is indistinguishable from any other blue marble), a random selection mechanism would choose a red marble with probability 1/10. Note that a random selection mechanism that selected 10 marbles from this bowl would not necessarily result in 1 red and 9 blue. In situations where a population consists of items that are distinguishable, a random selection mechanism requires equal probabilities for any item to be chosen. That is if the selection process is such that each member of a population, of say research subjects, has the same probability of being chosen then we can say the selection process is random. (*Wikipedia*)

    One of the most difficult tasks in science and almost all human efforts to find truth is to establish cause and effect: Was the outcome or change truly random? Did we find the true cause? Have we ruled out all the other possibilities or influences that created the outcome? In the example of the marbles described above, note some of the other factors that may affect the outcome: the selection would depend on the initial mixture, the time shaken, who shakes, who picks, who developed the mechanism, etc. And who would develop the "random selection mechanism"? What criteria would be used?

    And we must also keep in mind the following: When anything changes, for any reason, including chance and an action of free will, it almost always causes other changes, multiple changes, or a cascade of changes. There must be some organizing or coordinating principle, or else chaos.

Chapter Nine

Determinism

Condensed: This chapter is a follow-up on the previous chapter on chance and probability. It defines the concept of randomness and how it is used by scientists. One must understand the relationship between cause and effect to fully understand Darwin's theory of evolution.

*****

Determinism: *Determinism is the philosophical theory that all events, including moral choices, are completely determined by previously existing causes. Determinism is usually understood to preclude free will because it entails that humans cannot act otherwise than they do. The theory holds that the universe is utterly rational because complete knowledge of any given situation assures that unerring knowledge of its future is also possible. (Wikipedia)*

(Author's comments—the author recognizes that there are conflicting and more complicated definitions of the word *determinism* as used in this chapter. Among these are "pre-determinism," "soft determinism," "adequate determinism," and "indeterministic as defined by quantum physics." For simplicity and purposes of this discussion, the *Wikipedia* definition with the clarification as defined below will be used.)

The author's definitions as used in this chapter.

Creativity: The arts. Writing. Sculptor. Comedy and jokes, etc. The creation of something different—an original.

Determinism: As defined above, using the phrase "previously existing causes" to mean all causes originating with the big bang.

Freedom: With determinism, there is no such thing as free choice. In general, we think of free choice as the ability to make moral and ethical choices, as well as the routine decisions of everyday life, and to make those decisions strictly by ourselves, realizing that we are influenced by environment, heredity, experiences, etc. But always believing that humans being possess the power within themselves to choose between alternatives.

Free will: There can be no free will, as we usually use that phrase, in any system that has been determined.

Individuality: In strict determinism, there is no individuality. There would be no need for individuality in the Darwinian progression of evolution by natural selection. There are no individual roaches, and they have survived for millennia. Free will in itself means each human is an individual and capable of making decisions completely on his own.

Originality: Ideas. To be original, it must come out of nowhere. Determinism says that's impossible. Is what I'm writing an original thought, or was it determined at the time of the big bang?

The classical description of Determinism is that all things have been determined, from the earliest beginning at the time of the big bang. It is essentially an all or nothing proposition: not only just the natural laws that control the physical universe but every change, every thought by human beings, every action by plant or animal, has already been determined—that is, there is no human free will, originality of thought or action, no change in the physical world, the classical example of the wind-up clock. The universe came into being, and every action progresses along following fixed and irreversible principles. My writing this essay, your reading this essay, a bird singing in a tree—all inevitably resulted from actions occurring at the time of the big bang. The bird sits on a limb and looks around. Is he really free? Is he enjoying the view, or is he completely responded to instinct and looking for food or his mate? Or in neutral? Was it all determined?

DETERMINISM

Strict determinism theoretically rules out any possibility of chance, human free will, or an Outside Force. The word *chance* implies that something happened for no reason. Human free will means human beings can make decisions and originate ideas freely, not constrained by any preordained set of inviolable rules. An Outside Force, as used in this discussion, means "a force or being that operates outside of the natural laws and influences changes in the natural laws, as well as human behavior."

Any discussion of determinism must include an understanding of that phenomenon which we call chance. By definition, and when used properly, *chance* means "something that happens unpredictably without *discernible* human intention or *observable* cause." Most of us, somewhat glibly, use the word *chance* as if something happened for no reason whatsoever—there was no cause at all. It just happened. For most of us in common conversation, this is no problem, and somewhat meaningless. The strict determinists believe that every change has a specific cause and that all causes were created at the time of the big bang, but most are not discernible or observable. This interpretation leaves no room for changes due to chance or any other reason. At face value, this is the easiest and most comfortable of all the beliefs about chance and determinism.

But for those who do not accept determinism, a whole range of options become available to them. Most of what I have written in this treatise is based on the belief and philosophy that rejects the philosophy of determinism, as it reduces human beings to nothing more than cogs in the great machinery of the universe.

Chapter Ten

Consciousness

Condensed: This chapter is presented as additional evidence as to the complexity of the human brain as well as our limited ability to understand our own brains. That the human brain, with its ability to be conscious of itself along with the multitude of other abilities, is a product of random changes with no specific cause is well beyond the standard and accepted uses of chance and randomness by the scientists.

*****

Consciousness is the most advanced state in the development of the human brain. The efforts to understand that brain function that we call "consciousness" has challenged, stymied, and frustrated many of our brightest scientists and philosophers through the ages. The problem basically is that we have a situation in which "consciousness" is trying to understand "consciousness," an entity trying to understand itself. That sentence, as confusing as it sounds, emphasizes the difficulty in trying to understand and define this subject. It is a study unto itself, and we will not go into any analysis or detail in this treatise. The introductory section from the online *Wikipedia* is reproduced below to give the reader some idea of the present thoughts and understanding of consciousness. Anyone who has tried to understand consciousness immediately realizes the difficulty or impossibility of adequately understanding this subject.

CONSCIOUSNESS

This chapter is added to support two of the main points in this treatise: first, that the human brain has limits and reaches an endpoint in its ability to understand certain things, and secondly, all efforts so far to produce an "artificial intelligence" makes no attempt to program into their electronic devices anything resembling human brain consciousness. To claim that a robot without consciousness is the equal of the human brain seems ludicrous to me. (Whether this proves to be of importance is another subject.)

I offer the following definition and discussion to illustrate just how difficult it is to define consciousness.

> *Consciousness is the state or quality of awareness, or, of being aware of an external object or something within oneself. It has been defined variously in terms of sentience, awareness, qualia, subjectivity, the ability to experience or to feel, wakefulness, having a sense of selfhood or soul, the fact that there is something "that it is like" to "have" or "be" it, and the executive control system of the mind. In contemporary philosophy its definition is often hinted at via the logical possibility of its absence, the philosophical zombie, which is defined as a being whose behavior and function are identical to one's own yet there is "no-one in there" experiencing it.*
>
> *Despite the difficulty in definition, many philosophers believe that there is a broadly shared underlying intuition about what consciousness is. As Max Velmans and Susan Schneider wrote in The Blackwell Companion to Consciousness: "Anything that we are aware of at a given moment forms part of our consciousness, making conscious experience at once the most familiar and most mysterious aspect of our lives."*
>
> *Western philosophers, since the time of Descartes and Locke, have struggled to comprehend the nature of consciousness and identify its essential properties.*

Issues of concern in the philosophy of consciousness include whether the concept is fundamentally coherent; whether consciousness can ever be explained mechanistically; whether non-human consciousness exists and if so how can it be recognized; how consciousness relates to language; whether consciousness can be understood in a way that does not require a dualistic distinction between mental and physical states or properties; and whether it may ever be possible for computing machines like computers or robots to be conscious, a topic studied in the field of artificial intelligence.

Thanks to developments in technology over the past few decades, consciousness has become a significant topic of interdisciplinary research in cognitive science, with significant contributions from fields such as psychology, anthropology, neuropsychology, and neuroscience. The primary focus is on understanding what it means biologically and psychologically for information to be present in consciousness—that is, on determining the neural and psychological correlates of consciousness. The majority of experimental studies assess consciousness in humans by asking subjects for a verbal report of their experiences (e.g., "tell me if you notice anything when I do this"). Issues of interest include phenomena such as subliminal perception, blindsight, denial of impairment, and altered states of consciousness produced by alcohol and other drugs, or spiritual or meditative techniques.

In medicine, consciousness is assessed by observing a patient's arousal, and responsiveness can be seen as a continuum of states ranging from full alertness and comprehension, through disorientation, delirium, loss of meaningful communication, and finally loss of movement in response to painful stimuli. Issues of practical concern include how the presence of consciousness can be assessed in severely ill, comatose, or anesthetized people, and how to treat conditions in which consciousness is impaired or disrupted. (Wikipedia)

Author's comments—after reading the above, I think the reader will understand the difficulty of explaining and defining human consciousness.

Chapter Eleven
The Theory of Evolution

The question is not which came first: the chicken or the egg? The question is: where did the chicken or where did the egg come from?
—C. S. Lewis

Condensed: This chapter presents a general overview of the development of the theory of evolution of the universe we live in as we understand it. It specifically discusses the understanding of Darwin's theory of human evolution as generally accepted by scientists and laypeople today. The author makes clear that this is a limited and rudimentary presentation of a very complex subject, what with the massive amount of material and ideas that have been published before and after the works of Charles Darwin.

*****

A concise description of the latest scientific thinking on the theory of evolution taken from the *Encyclopedia Britannica*:

> *The theory in biology postulating that the various types of plants, animals, and other living things on Earth have their origin in other preexisting types and that the distinguishable differences are due to modifications in successive generations. The theory of evolution is one of the fundamental keystones*

of modern biological theory. The diversity of the living world is staggering. More than 2 million existing species of organisms have been named and described; many more remain to be discovered—from 10 million to 30 million, according to some estimates. What is impressive is not just the numbers but also the incredible heterogeneity in size, shape, and way of life—from lowly bacteria, measuring less than a thousandth of a millimetre in diameter, to stately sequoias, rising 100 metres (300 feet) above the ground and weighing several thousand tons; from bacteria living in hot springs at temperatures near the boiling point of water to fungi and algae thriving on the ice masses of Antarctica and in saline pools at −23 °C (−9 °F); and from giant tube worms discovered living near hydrothermal vents on the dark ocean floor to spiders and larkspur plants existing on the slopes of Mount Everest more than 6,000 metres (19,700 feet) above sea level.

The virtually infinite variations on life are the fruit of the evolutionary process. All living creatures are related by descent from common ancestors. Humans and other mammals descend from shrew-like creatures that lived more than 150 million years ago; mammals, birds, reptiles, amphibians, and fishes share as ancestors aquatic worms that lived 600 million years ago; and all plants and animals derive from bacteria-like microorganisms that originated more than 3 billion years ago. Biological evolution is a process of descent with modification. Lineages of organisms change through generations; diversity arises because the lineages that descend from common ancestors diverge through time.

The 19th-century English naturalist Charles Darwin argued that organisms come about by evolution, and he provided a scientific explanation, *essentially correct but incomplete, of how evolution occurs and why it is that organisms have features—such as wings, eyes, and kidneys—clearly structured to serve specific functions. Natural selection was the fundamental concept in his explanation. Natural selection occurs because individuals having more-useful traits, such as more-acute vision or swifter legs, survive better and produce more progeny than individuals with less-favorable traits. Genetics, a science born in the 20th century, reveals in detail how natural selection works and led to the development of the modern theory of evolution. Beginning in the 1960s, a related scientific discipline, molecular biology, enormously advanced knowledge of biological evolution and made it possible to investigate detailed problems that had seemed completely out of reach only a short time previously—for example, how similar the genes of humans and chimpanzees might be (they differ in about 1–2 percent of the units that make up the genes).*

Anyone superficially perusing or doing an in-depth study of the available literature and opinions on Charles Darwin's theory of evolution is immediately confronted with an overwhelming task. The materials, written and now available in electronic forms and videos, predate Charles Darwin's two great scientific presentations in his *Origin of Species* and *The Descent of Man*. The massive amount of information includes, besides the works of Darwin himself, research by thousands of scientists, and theories, ideas, and opinions of people from almost every walk of life. These include philosophers; the extremes of the seminarians from total acceptance of Darwin's theories to those proclaiming that Darwin was sent from the devil; politicians; and laypeople of every persuasion expressing their opinions. Even at the present time, there continues to be significant disagreement about the evolution of human beings, even among the best read and most studious of those trying to understand the full story of that evolution.

This treatise makes no effort to present an in-depth presentation of all the multiple facets of this challenging and controversial subject. By tradition, most of us understand that science is concerned with the "how and what" of the changes wrought by evolution, sometimes asking "why?" but only when the *why* helps them understand the *how*. Scientists, by definition, stop short of going beyond any quest for a cause after they are satisfied that they have proven their theory. Science does not look for a Primary Mover. The main point of this treatise is to ask *why* the evolutionary changes happened. This work tries to present a coherent analysis that challenges the accepted theory that the changes occur due to "chance mutations," that nothing ever happens without a specific cause, and that there has to be some "Outside Force" that causes these changes. The preceding pages of this book were all presented in order to focus on the cause of human evolution.

The general evolution of the universe to its present status as we understand it is as follows:

- ❖ All the energy of the universe is condensed and exists at one concrete point in time.
- ❖ There is suddenly a "big bang," and all this energy expands in all directions.
- ❖ Over billions of years, some of this energy transforms into mass.
- ❖ The mass evolves into elements.
- ❖ The elements condense into stars and planets.
- ❖ The planet Earth is a part of this evolution.
- ❖ Over billions of years, the Earth cools, and the various elements that make up the Earth evolve into a myriad of solids and liquids, and over time become basically what the planet Earth is now.
- ❖ After this initial phase, the "primordial ooze," a large mass of water filled with the elements that have formed develops and covers a large part of the planet. The billions of

elements that are floating around in the big ooze are constantly colliding with each other and develop into molecules. These molecules, constantly colliding with each other, form complex entities. These entities are constantly changing and begin to form tissues and organs, which in time, evolve into complex "living" plants and animals.

❖ Some of the animals become amphibious and move out of the water onto the dry land and further evolve into more complex entities. *It is at this point that all the controversies about evolution begin and the theories for the causes of the evolutionary processes develop.* In the mid-1800s, Charles Darwin, combining some previous theories on human evolution with his in-depth studies of fossils and modern fish and animal life in the Galapagos Islands, writes his books. His first book, *On the Origin of Species*, published in 1859, outlined his theory of evolution: all species randomly mutate and those that change into a form capable of surviving the changing environments survive. *On the Origin of Species* does not include an opinion on human beings. His second book, *The Descent of Man*, published in 1871, presents Darwin's theory of the evolution of human beings. His theories are based completely on *chance* mutations and survival of the fittest.

❖ Darwin's books create a firestorm of critics and supporters. That human beings developed from single-cell organisms, through multiple stages, many of which are closely related to, or descended from, other animals challenges the basic beliefs of a major portion, if not the majority, of the people of his day and into present times. The Jewish and Christian faiths, then and now, hold fast to the belief that the planet Earth and human beings in their present forms were formed by a Creator God. Many believe that this creation took place in in a matter of days in the very recent past. Others believe that the planet and human beings evolved according to the Creator's master plan.

THE THEORY OF EVOLUTION

The scientists themselves, then and now, are divided. Some believe that Darwin's theories, supported by new developments in understanding of genetics and molecular biology, prove that humans and animals evolved from one stage to the next purely by chance, and the lives of human beings are purely materialistic. Others believe that the evidence supports the evolutionary processes as outlined and that the changes can be defined by chance and probability looked at from a purely objective view. But that does not rule out other reasons for the changes, including some Outside Force that we do not understand.

The teaching of evolution in public schools continues to be controversial. One side feels strongly that only the accepted scientific presentation should be taught, and any variations from this position are considered religious and violate the laws regarding separation of church and state. Parents, churches, and parochial schools can teach their students other theories or modifications.

The other side feels that presenting only the scientific understanding of evolution, without the presentation of other possibilities, biases the students in the public schools. Many believe that any presentation of the evolution of human beings involves some degree relative to the value of human beings in the overall scheme of the universe. To teach only the accepted scientific position, which is basically materialistic, without making it clear to the students that there are other alternatives or modifications, is, in itself, the teaching of a religion.

There are many objections and challenges to the theory of evolution. Some of these challenges are presented in this chapter. The author's motive is to present this specific case: the scientific theories make no effort to explain the causes of evolution other than by chance. The point is that this is satisfactory for the scientist, and should be. But the layman should understand that this is not the end the story. The use of "chance" and "probabilities" is a standard and useful scientific tool, but in no way speaks to what "causes" the phenomenon they are studying. The possibility of an Outside Force causing the changes is not negated by the scientific method. In fact, by definition, by using chance or probabilities to support a theory, the scientist admits that there is still "wiggle room" for some other explanation. A very simple example is when mathematicians or engineers use pi (π). π is an irrational number, and when used in any equation or calculation, it means that the final result will never be 100 percent accurate. But it serves its purpose to support the final conclusion pragmatically. The bomb went off; the bridge still stands.

One of the major objections to the theory is loosely stated: the assumption that given enough time, anything can happen. This is not meant cynically. All supporters of Darwin's theory accept as fact that the evolution of *Homo sapiens* on the planet Earth required millions of years to reach its present status. The point is that, as the mutations occur, leading one species to better survive its environment, that environment drastically changes in the time frame. As soon as the species adapt to its present environment, that environment changes, requiring the species to mutate again, almost *ad infinitum.*

Note that Darwin himself wrote, "adaptability is more important than survival of the fittest," but in most discussions, natural selection and random mutations are synonymous with Darwin's theory of evolution. Darwin's theory of evolution differs from the other sciences in that it is, inevitably and unavoidably, associated with the beginning of human life, and so inevitably and unavoidably challenges or refutes many of the basic long-held beliefs of the religious and the philosophers. And the theory of evolution by mutations and natural selection is burdened by incredible generalizations and assumptions but is accepted by most because no better theory has been proposed. After all this time, one would think there would be multiple theories. Maybe it's because the answer lies closer to the possibility of an Outside Force.

One of the most important components necessary to make Darwin's theory of evolution work is coordination. The multiple changes resulting from transmutation, at some point, have to be coordinated or there is chaos. And of course, survival of the fittest doesn't necessarily mean progress, progress in the sense of advancement or improvement. The entity that survives in a harsh environment does not automatically assure an improvement or movement toward a "better" product. *Homo sapiens* is presented as the most advanced living organism, and its species has resulted from random mutations and natural selection. For such a highly organized body, with its brain, to be a result of pure randomness without some coordination or plan goes beyond any of the scientific uses of probabilities. Whether one uses the word *God, Outside Force, pressure*, or some other nonspecific, generic term, there has to be "something else." There is no denying the fact that the human "brain," with few exceptions, throughout the long history of human beings on this planet, has felt a need for an explanation for these causes. Is this need a result of chance mutations? If so, how does it add to the chances for survival?

One of the greatest challenges to the Darwinian theory is this: any major change requires other changes, frequently multiple changes. For an example, the next chapter on "Strengths and Weaknesses" describes the processes required for the development of a longer neck in an animal, the giraffe.

In Darwin's time, many believed that the human body had reached its maximum point of evolution but would continue to evolve emotionally, intellectually, spiritually, and this evolution would result in a perfect human species in which all the unacceptable traits would have been weeded out, and utopia would have arrived. But *Homo sapiens* developed a conscience, a feeling of sympathy, which, up until this point, was a positive trait aiding the survival of the species. This in turn produced another concern, what is generally referred to as Social Darwinism. Survival of the fittest means, by definition, the unfit do not survive. The weakest cannot care for themselves and die out, leaving only the strongest.

At this point in the evolutionary struggle, the sympathetic trait of the human brain led humans to help the unfit survive and continue to multiply. The evolutionary process is slowed, stopped, or even reversed. Is survival of the fittest a self-limited process in *Homo sapiens*? Or can *Homo sapiens* develop societies that can progress carrying the baggage of those too weak to provide for themselves? If so, will this be the next step in the evolutionary process, two separate species: *Homo sapiens weak* and *Homo sapiens strong*? Or will there be two societies: *Homo sapiens weak* and *Homo sapiens strong*?

If a species cannot survive, why rescue "endangered species" now? If humans are superior or more fit, then too bad for the snail darter. Are we upsetting the progressive evolution? If man interferes, then we get a different outcome. Does this not negate or seriously interfere with Darwin's basic reasons for evolution? If you believe in Darwin's theory of evolution, then you cannot believe in Social Darwinism—that is, interfere and help the weak survive. If you do believe in Social Darwinism, then you must have a reason for humans to interfere. You can say, humans have developed (evolved to) a feeling toward others that makes them "better" and allowed to make these decisions. But who's to say we're better?

Any definition of the word *evolution*, whether referring to Darwinian evolution or any other type of evolution, means, as defined by *Merriam-Webster*, "*a process of continuous change from a lower, simpler, or worse to a higher, more complex, or better state.*" The universe evolved. Human evolution was part and parcel of the evolution of the universe. From the middle to the late nineteenth century, evolution was generally accepted in universities, mainly on the idea of transmutation—one species evolved into another through natural, not supernatural, mechanisms. Survival of the fittest was accepted by very few scholars. In the twenty-first century, some scholars accept survival of the fittest as part of the evolutionary process; some do not. The understanding of evolution itself has evolved. Evolution means change. The universe itself is evolving and changing. So human beings are part of the universe that is changing.

Summarizing this chapter—there is no denying that the universe and we human beings evolved from simple structures to our present selves. The scientific method using probabilities has produced a reasonably accurate picture of the development of the natural world from the time of the big bang to our present day. The study of fossils and the ability to establish accurate periods of time by carbon dating have presented an almost-irrefutable case for the evolution of the human body. These methods explain what happened. But these methods do not and cannot adequately explain the causes of the changes. Chance, randomness, and the use of probabilities have proved useful and extremely helpful in regards to predictability and understanding, but by definition no attempt to determine *causes*. The point here is that the scientists realize the limitations of this method, but most of our institutions and presentations of the theory of the evolution of the human body has failed to make this clear.

Chapter Twelve

Strengths, Weaknesses, and Assumptions

Condensed: This chapter presents some of the strengths and weaknesses of Charles Darwin's theories. It is by no means complete, and many of its points are debatable and clearly can be challenged. The point remains, however, that there are many legitimate strengths and weaknesses of his theories, and every student of evolution should be aware of these challenges.

*****

The positives and negatives of Darwin's theory of evolution have been debated vigorously by thousands of scholars for over 150 years. In my studies for this project, I have noted a very strong bias: a heavy preponderance of the articles and studies compare the accepted theory of evolution with the traditional creationist's position. It appears that the Darwinists have selected an easy opponent to debate in a forum that is overwhelming suited to their type of discussion (objectivity).

With great understatement, it is obvious that this study is not complete enough to evaluate all the strengths and weaknesses of the theory. I refer the reader to any of the multiple sources available to discuss this issue. The remainder of this chapter is my list of questions and concerns that I feel are not adequately explained by Darwin's theory.

STRENGTHS, WEAKNESSES, AND ASSUMPTIONS

First and foremost, among the weaknesses is this: no plausible explanation for the incredible coordination required to bring about the improved and advanced status of the next stage in an evolutionary process.

The development of "the useless manifestations"—that is, useless manifestations in the sense of "not absolutely necessary to maintain or defend life." Why did the single mutation toward spirituality survive in a world in which it was survival of the fittest and not survival by an isolated physically weak trait? In genetics, selection of one trait usually comes at the expense of another. This would affect every case of survival.

Everybody is different: every human face is different and recognizable, there are multiple body sizes, different physical traits and abilities, as well as mental, and a host of other differences. That is why we cannot define a common human being. Question—Why has evolution not produced more common individuals? Why such differences? The human face? Why the weak and the strong? How does this fit in with survival of the fittest?

Look at the following example: The development of the human and animal eye is commonly used as an example of mutations that develop into a complex organ. Darwin's theory of evolution says that at first there was the development of a cell or a cluster of cells that were sensitive to light and responded with certain actions. These became tissues and then organs and over a long period of time became a functioning eye. It makes no plausible argument as to the development of the parts of the eye that had nothing to do with light sensitivity—that is, the lens, the cornea, blood flow, protective tears, etc. And most important, it gives no explanation for the development of the optic nerve which in turn ends in the brain, and no explanation for the brain's ability to understand what it "sees." It makes no explanation for the fluids that give the eye form, the formation of tears to protect the cornea, etc. And then, of course, there is the essential requirement for blood vessels which carry oxygen and nutrients to every cell in the eye and remove waste products. And of course, the blood vessels work in combination with the heart, lungs, liver, and kidneys. Again, no explanation for the coordination.

George Gallup, the noted pollster and mathematician, has stated that the odds that all of the above factors developing into one organized, functional organ is on the order of 10 to the power of thousands.

Another example—the giraffe's neck. Lamarck postulated that giraffes increased the length of their necks by stretching their necks to reach for higher food. This acquired trait would be inheritable, leading to the evolution of a new species. But Lamarck's theories are not accepted by modern scientists. This actually makes more sense to me than chance mutations and survival, because a gradual stretching of the neck would involve an effect on all the organs and structures involved, including the heart and lungs, and is conceivable, somewhat like bodybuilding, when increased muscle mass requires increased cardiac output.

Darwin's theory of evolution says that the giraffe evolved when food supplies ran out on the lower levels. The small animals needed food on the higher levels that they could not reach, so by chance, an individual was born with a longer neck, and this started a new generation of long-necked animals. This in no way explains the processes required to result in a longer neck. There must be replication of vertebral bodies, lengthening of arteries, veins, nerves, the esophagus, the spinal cord, etc. There also must be enlargement and strengthening of the heart in order to pump blood to the head at this additional height. There also must be simultaneous development of compensatory mechanisms to adjust changes in blood pressure when the head is down versus when the head is up. All these changes require additional energy, which requires additional food, which was the problem to begin with. It would also require hundreds of years, during which time the environment would change radically and a longer neck would not necessarily be in advancement.

Another example would be to describe the development of blood clotting, or coagulation. The following illustration and paragraph describe the very complex process as outlined in *Wikipedia*. The layperson can peruse through this process without a complete understanding, but will see that a highly complicated *series* of events has to happen in order for blood to clot.

STRENGTHS, WEAKNESSES, AND ASSUMPTIONS

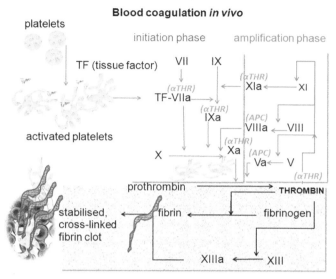

Coagulation begins almost instantly after an injury to the blood vessel has damaged the lining of the blood vessel. This initiates two processes: changes in platelets, and the exposure of subendothelial tissue factor to plasma Factor VII, which ultimately leads to fibrin formation. Platelets immediately form a plug at the site of injury; this is called primary hemostasis. Secondary hemostasis occurs simultaneously: Additional coagulation factors or clotting factors beyond Factor VII respond in a complex cascade to form fibrin strands, which strengthen the platelet plug. (Wikipedia)

If one of the processes fails to produce the next stage, then the cascade stops and will fail to complete the clot. Note—the organism's survival depends on the completion of *all* the stages. Let's look at an organism that is in the first stages of development according to "survival of the fittest" theory. The organism is injured. Without the *total* clotting mechanism, the organism cannot survive and the evolutionary process stops at this point. To counter this analysis, the Darwinist would have to claim that the entire clotting mechanism evolved in one generation. It is difficult to imagine how such a complex and necessary part of survival could develop purely randomly, even in millions of years, let alone one quick stage.

THE DEATH OF ANNIE

In other words, there must be a high degree of coordination at all levels. Darwin's theory of evolution does not attempt to explain how all the mutations that occur are coordinated to form an advanced organism with all its organs and tissues dependent on each other. For example, look at one organ: the kidney. The kidneys function as a filter and create a liquid waste product. This product must be excreted to the outside of the animal. A tube was formed for this purpose. Then a bladder was formed to store this liquid prior to its excretion outside the body. This required a valve and a controlling mechanism. These developments had to be coordinated with each other. Every organ in the animal bodies requires this sort of coordination. It takes an extremely liberal interpretation of the laws of probability to defend the position that these changes occurred without some outside pressure.

Why has Darwin's theory of evolution not added to the longevity of man? Human beings still age at the same rate as they did thousands of years ago. Note the average has increased but not the maximum. The sturdiest human bodies that are not destroyed by diseases or accidents still disintegrate and die short of one hundred years, with most dying prior to seventy years. Can the Darwinist make a legitimate claim that too short a time has passed to affect human longevity?

Not every evolutionary change is for the betterment of the species. Would it not be better if mammals retained their ability to breathe in water as well as on land? How about the difficulty for the warm-blooded animals to adapt to changes in temperature?

Darwin and his followers constantly use the phrase "nature selects" or "natural selection." Who or what are they referring to by the term "nature"? *Merriam-Webster* defines nature as, *"the external world in its entirety. Humankind's original condition." Dictionary.com* defines nature as, *"the material world, especially as surrounding humankind and existing independently of human activities. The natural world as it exists without human beings or civilization."* Natural selection, of course, means selection by "nature." By any interpretation, the evolutionist's use of "nature or natural selection" refers to some nebulous, something out there causing a change in itself. This apparent contradiction remains a useful strictly scientific explanation but again fails to address the question of what is on "the other side of nature," the primary cause.

Darwin's theory of evolution makes the point that everything is related to survival—that is, all changes are based on mutations that lead to one species surviving and one species dying. So many human traits—selfishness, laziness, cowardice—are meaningless or counterproductive. This does not explain why any of these traits, unrelated to survival, developed or especially persist. How explain the development of a conscience? You certainly can make a strong case that a conscience is counterproductive to survival.

There are so many unanswered questions. Darwin's theory of evolution answers only a small percentage of the questions regarding the development of the human body, especially the brain.

And emotions, are they necessary for survival? Can they be counterproductive for survival? Emotions are so strong that they frequently lead to the opposite of survival.

If Darwin's theory of evolution says that human beings evolved by mutations, then why are not all human beings exactly alike? If a single mutation brings about a single positive change, where did all the multiple changes in characteristics of humans come from? Why are some more capable of survival than others?

Darwin's theory of evolution has no ideas on the persistent, through the centuries up to the present time, of human beings never-ending quest to understand the transcendent things. These feelings are in almost every human being and just won't go away. If at all, they play almost no part in survival. Indeed, the argument could be made that they impede survival: transcendent beliefs lead human beings to do things that hurt their chances of survival and may even lead to their death.

If a species develops because of a random mutation and then survives, and then this mutation is transmitted to the next generation, then the next generation should be an exact replica. If more changes than the random mutation have developed, then what force determines whether improved changes are passed on? If the new generation is an improvement on the old, then why does not some species, especially humans, each succeeding generation be a cumulative improvement? That is, with humans, why not pass on memories, experiences, physical improvements, etc. and develop a super species?

There are too many human traits that have nothing to do with survival. To name a few—moods, sighing, closing your eyes and enjoying music, the very fact of enjoyment of anything, walking slowly when you feel bad and with a fast gait when you feel good, etc.—thousands of things that are a part of basic human behavior in life that have no connection to survival or chance mutations.

Why the strong feelings one sex has for a singular individual of the opposite sex—that is, what we call being "in love"? These feelings inevitably lead to a situation where there are fewer offspring as opposed to *carte blanche*, or "free-love," sexual reproduction, and therefore less chance of survival of the species. Without these feelings, humans would certainly produce considerably more offspring and increase the survival rate.

Why did humans continue evolution in the spiritual, moral, emotional, arts, etc. after his animal body was basically complete?

Why do humans value life so much—take care of orphans, the injured, the infirm, the aged, and those who have no positive effect on survival? Indeed, if survival of the fittest is the goal, these values help achieve just the opposite.

The problem is that those who do not believe in transcendent causes offer no, or at best inadequate, explanations for the multitude of questions that humans have asked through the ages and continue to ask in the present.

Here follows a short list of relative and nonspecific words and phrases frequently used by Charles Darwin himself, as well as other modern writers. The literature on this subject is filled with similar words and phrases. This list is just a small sampling. Rarely, if ever, in the scientific literature does one see words or phrases like "it appears," "unexplained," "inevitable, possibly," "variability is governed by many unknown laws," "blind chance," "random changes," "may or may not help," "the origin of life requires some raw material that could allow the spark of life to emerge"...

STRENGTHS, WEAKNESSES, AND ASSUMPTIONS

From Darwin's *Dissent of Man*—"probably," "may happen," "it happens," "occasionally," "possibly," "random," "it happened," "might," "impossible," "improbable," "unless," "it appears," "a tendency," "inevitable," "must have," "assume," "unexplained," "hardly."

From Darwin's *Origin of Species*—"appears occasionally," "changed conditions," "variability is governed by many unknown laws," "appears to have played," "has no doubt largely aided," "the cultivator may have disregard," "selection seems to have been the predominant power."

From Paul Davies's *God and the New Physics*—"simultaneously organized," "blind chance," "arose spontaneously," "small coincidences are relatively much more likely than big ones."

From Bill Nye's *Undeniable*—"it happened," "came through," "mutated," "something would have to happen," "a favorable set of mutations came through," "no one is absolutely sure," "it may provide," "random changes may or may not help," "in those lots are bound to be new configurations," "had to infer," "the origin of life requires some raw material that could allow the spark of life to emerge," "it has a chance," "something would have to happen to us to create a new species," "and probably up through," "was an apparent result," "we would expect," "it may be that nature allows."

From *Scientific American Magazine*, June 2018, "How Did Life Begin"—"condensed," "stuck together," "it was not too hot…not too cold," "probably volcanic," "were likely," "nudged along," "eventually formed," "became enclosed," "spontaneous assembly."

And I would add the following reference from Lecompte du Noüy's *Human Destiny*, on his challenge to Darwin's theory:

> *In brief, each group, order, or family seems to be born suddenly and we hardly ever find the forms which link them to the preceding strain. When we discover them they are already completely differentiated. Not only do we find practically no transitional forms, but in general it is impossible to authentically connect a new group with an ancient one. Therefore, the problem remains as to whether the passage was*

more or less sudden or more or less continuous. As we have seen, probability clearly indicates that only groups which have existed long enough to multiply and become widely dispersed can be found in the fossil stage. There is, then, nothing astonishing in the fact that we do not find initial forms. These observations lead to important conclusions which do not seem to have been put in evidence so far—namely that the transitional forms are not stable forms; they do not multiply in great numbers and do not propagate. They have another role to play. Once more, everything takes place as though there were a goal to be attained, a highest stage of development destined to evolve still further; as if the intermediary form lost its importance the moment the next stage had been started. A relationship exists between the two stages similar to that which connects an industrial prototype with the manufactured product, on condition that the last prototype incorporates an improvement, absent in all the preceding ones, which confers a superiority worthy of being tried out on a large scale. In the case of natural evolution this untried character must be hereditary.

Chapter Thirteen
Science

> *The most beautiful and profound emotion we can experience is the sensation of the mystical. It is the sower of all science. The one to whom this emotion is a stranger, who can no longer wonder and stand rapt in awe, is as good as dead. To know what is impenetrable to us really exists, manifesting itself as the highest wisdom and the most radiant beauty which our dull faculties can comprehend only in their most primitive forms—this knowledge, this feeling is at the center of religiousness.*
> —Albert Einstein

Condensed: To be fully aware of the complex changes and challenges of modern life in the twenty-first century, every person must have some rudimentary grasp of science in general, its abilities and limitations and dangers. This chapter is an obviously limited presentation of one of the most complex and difficult-to-grasp and fully understand part of human existence as we find it. To fully understand Darwin's theory of human evolution, one must understand the basic methods of the scientific approach to the search for truth. Darwin's theory is a *scientific* product, not philosophical or metaphysical.

*****

THE DEATH OF ANNIE

It is important in any discussion of science to understand the meaning of the words and phrases that are used. Listed below, in alphabetical order, are the definitions of some of the important words and phrases that are commonly used, with the author's comments as to the way they are used in this treatise.

Bias (*Dictionary.com*): a systematic as opposed to a random distortion of a statistic as a result of sampling procedure.

(*Science Daily*): In psychology and cognitive science, confirmation bias (or confirmatory bias) is a tendency to search for or interpret information in a way that confirms one's preconceptions, leading to statistical errors.

Hypothesis (*Merriam-Webster*): something not proved but assumed to be true for purposes of argument or further study or investigation. A proposition tentatively assumed in order to draw out its logical or empirical consequences and test its consistency with facts that are known or may be determined.

Intuition (*Dictionary.com*): a phenomenon of the mind, describes the ability to acquire knowledge without inference or the use of reason. The word intuition comes from Latin verb *intueri* translated from the late Middle English word *intuit*—to contemplate. Intuition is often interpreted with varied meaning from intuition being glimpses of greater knowledge to only a function of mind; however, processes by which and why they happen typically remain mostly unknown to the thinker, as opposed to the view of rational thinking.

Natural world (*Merriam-Webster*): all of the animals, plants, and other things existing in nature and not made or caused by people. [The author's definition and as used in this essay/book—the universe, but more specifically the study of matter and all things physical, as opposed to the transcendental, those things which are not matter or physical.]

Science (*Wikipedia*): (from Latin *Scientia*, meaning "knowledge") a systematic enterprise that builds and organizes knowledge in the form of testable explanations and predictions about the universe. In an older and closely related meaning, "science" also refers to this body of knowledge itself, of the type that can be rationally explained and reliably applied. Ever since classical antiquity, science as a type of knowledge has been closely linked to philosophy. In the West during the early modern period the words "science" and "philosophy of nature" were sometimes used interchangeably, and until the 19th-century natural philosophy (which is today called "natural science") was considered a branch of philosophy.

In modern usage, however, "science" most often refers to a *way of pursuing knowledge,* not only the knowledge itself. It is also often restricted to those branches of study that seek to explain the phenomena of the material universe. In the 17th and 18th centuries scientists increasingly sought to formulate knowledge in terms of laws of nature. Over the course of the 19th century, the word "science" became increasingly associated with the scientific method itself, as a disciplined way to study the natural world, including physics, chemistry, geology, and biology. It is in the 19th century also that the term scientist began to be applied to those who sought knowledge and understanding of nature. However, "science" has also continued to be used in a broad sense to denote reliable and teachable knowledge about a topic, as reflected in modern terms like library science or computer science. This is also reflected in the names of some areas of academic study such as social science and political science.

Scientific method (Dictionary.com): a method of research in which a problem is identified, relevant data are gathered, a hypothesis is formulated from these data, and the hypothesis is empirically tested.

Second law of thermodynamics: the author's very loose interpretation—there is no such thing as a perfect steam engine, automobile, rocket, human body, etc. Every entity is inefficient to some degree and inevitably doomed to disorganization. Example, once an automobile leaves the assembly line (its point of maximum organization), it immediately begins the process of disorganization, and at some point in time will become totally disorganized.

Theory (Dictionary.com): a coherent group of tested general propositions, commonly regarded as correct, that can be used as principles of explanation and prediction for a class of phenomena. A proposed explanation whose status is still conjectural and subject to experimentation, in contrast to well-established propositions that are regarded as reporting matters of actual fact. [The author adds, "appears reasonable, logical, etc. but not proven; should be viewed as possible, or even probable, but always keep in mind may not be true."]

*T*his chapter is in no way an attempt to discredit or cast doubt on the scientific method, or on the advantages and great positives of pure objective reasoning and analysis. The following pages present a case for the positives of pure science while presenting a case for caution in accepting without reservation that the scientific method has always produced the truth or an understanding of reality. The history of science makes clear that what appeared to be obvious, turned out to be not just inaccurate, but on occasion, the exact opposite.

History includes a long line of scientists, humanists, philosophers, theologians, and writers and thinkers of every persuasion who have warned of the dangers of unrestricted scientific progress without regard to values. This last issue, as incredibly important to humanity as it is, is perhaps the most difficult to define and address. Not just because the rogues misuse the methods but people of good faith and the highest motives can inadvertently lead humanity down the road to Armageddon. We must always keep in mind many human activities have a negative side, and more often than not, the negatives appear only after it is too late.

The Nobel Peace Prize is given in the name of the man who invented dynamite. One of the most peaceful men in the twentieth century produced the basis for the creation of the nuclear bomb. Look at the negative side of the modern Internet and its effect on personal relationships and negative effects on younger and maturing children. Are these scientific evolutions an inevitable result of basic scientific research? Is there any conceivable way to anticipate the unknown and restrict or limit scientific research and progress? Should our educational and public institutions require the teaching of values, and the possibility of transcendental realities, that would guide the individuals who become the scientists and researchers?

In modern usage and in today's terms, what is a scientist? Is a scientist defined as someone who spends his professional time studying the natural world? If he has a BS in one of the natural sciences, does this make him a scientist? Does the scientist need to be board-certified? Does the scientist have to be purely objective about all things, or can he also have beliefs in transcendent realities? When someone claims to be a scientist, how can we know he is objective? Do you have to be a full-time researcher or teacher to claim the title of scientist? Is it accurate to say that a scientist is one who feels that if you can't prove it, then it is wrong or not the truth?

The scientist has chosen to study the natural world. But it appears to me that many scientists have "chosen sides"—that is, the natural world is the only area where truth can be found. These people seem to have dismissed as illegitimate any other types of thought or investigation. The transcendent world does not exist. Some seem to have lost the absolute basic motivation for investigation: the search for the truth, wherever that might lead them. By definition, the scientist is not interested in "primary" or "first" causes in the way philosophers or theologians are. The scientist looks back only as far as is necessary to find a cause and effect relationship. He is not interested in a "primary" cause, only a cause that satisfies his objective. The scientist, of course, as any other seeker of ultimate truth, may, outside of the laboratory, seek a primary mover.

Very importantly, the integrity of scientific research depends on the integrity of the individual researchers and writers. Faulty research, whether intentional or not, usually is corrected by the processes of science, because they prove unworkable or irreproducible. Peer review is a powerful instrument in helping scientists help each other in correcting biases and false conclusions. But this leaves a void where the researcher not committed to the highest standards of scientific ethics can mislead other researchers and the general public astray. We must always keep in mind that today's knowledge may not be the final word and always be open-minded, willing to change our beliefs and understanding with new knowledge, and never become passive or cynical. And if a scientist does not have a sense of integrity, what other controls are there? There are "bad "scientists just as there are "bad" doctors, philosophers, teachers, ministers, and a host of others who influence the population.

Having said all that, relative to the past, we live in the golden age of science. All the influences as mentioned above, in association with a highly developed sense of ethics regarding experimentation on human beings as well as animals, add to the immense respect scientists enjoy. By refusing to accept any new findings or data from unethical, closed societies who have no respect for human values, the scientist in the open societies do their part in upholding the dignity of humanity.

SCIENCE

All human beings, even those who try their best to be truthful and honest, have two problems: ego and bias. As Jane Austen said, *Pride and Prejudice*. Developing a hypothesis to fill a need or solve a problem is the easy part in most instances. Being totally objective and not being swayed by ego, or unable to see their own biases, challenges the most dedicated and committed researchers. And of course, desire for fame, money, or self-satisfaction is a powerful influence, along with the basic complexity of most scientific endeavors.

Note all the references to "scientist" in today's media and literature. He or she is a scientist. Therefore, their word is indisputable. Everybody used to settle an argument by saying "I read in the paper." Now we add to that "I heard it on the news." Or, more commonly, "I read it on the Internet." Of course, now most of us do not believe this. But many times we assume that if we read something that has been published by, or the word *scientist* has been attached, we give it credence, sometimes more than it deserves. My point is that in order for any one of us to be properly informed, we must always note the sources of the information that is presented to us. The true scientist is totally committed to determine the truth and accuracy of his project. The true scientist also knows the limits of any method devised by humans to determine the ultimate truth, including the scientific study of the natural world.

The scientist and all others interested in finding the truth must, in addition to being critical, unbiased, and committed to finding the truth, make every effort to ensure that the nonscientific audience understands the limits of the scientific method and the ever-present possibility of error or inaccuracy. Earlier scientists thought the world was flat, the Earth was the center of the universe, etc. Scientists in every era have been wrong, incomplete, and sometimes deliberately misleading.

My point is that the best and most dedicated scientists are often not only wrong but conclude after much study the exact opposite of the truth. There are many examples of the brightest and most committed scientists throughout history being wrong and in many cases convinced that the exact opposite is true. Pick any scientific text or reference book from one or two generations ago and note how many findings of that era have been modified or reversed.

Why are so many people today skeptical of many scientific claims? I think it's because of the misuse or careless use of statistics resulting in many erroneous conclusions. The competent scientist is well versed in the proper use of statistical methods and analysis of chance and random phenomena. Unfortunately, much information is presented to the public at large from the media and the academic world based on inaccurate and misuse of statistics, resulting in misunderstanding of the subject at hand. In the complicated and technical world in which we live, we must develop a better way of presenting scientific studies to the lay population in order to have a properly informed society.

I think it's also important to keep in mind that the scientific method has not been used in the development of many of the advancements in the history of humanity. For example, the development of automobiles, airplanes, computers, most aspects of space travel, etc. These things were developed through stages, but not randomly. These advances were made through free choice, jury-rigging, dreaming, etc. Note the absence of the scientific method in so many advancements in material and physical things. Note how many things have been developed by accident (discovery of penicillin) or a progression of events (someone notes that steam expands and causes motion and someone else notes this motion can be used to move a wheel, and someone else builds a steam engine and someone else puts it on a boat).

Sending a man to the moon and back was done mostly by engineers and jury-riggers (basically the same thing) and not merely following the rules of so-called scientific investigation. The modern computer developed through a multitude of steps, going from a need to crunch numbers to a teenager using a smartphone to text his girlfriend. Much progress has been motivated by need, and someone developed a product. Sometimes things just happen.

Have many scientists, educational institutions, and news media promoted science above its actual abilities and competence? Have they propagandized in their own way? Have they basically taught

SCIENCE

or tried to teach or inculcate in their students the idea that there is no possibility of transcendent things? Have they gone overboard in creating in their student's minds the idea "If you can't prove it, it doesn't exist"? And very importantly, are they saying that you have to use their method, the scientific method, and if you don't use their method, there is no way you can find the truth?

In the Middle Ages, the pope and the church told the scientists and the researchers what they could or could not print or teach. The scientists that believed or taught anything contrary to the teachings of the pope and the church would be excommunicated and lose their teaching positions, imprisoned and punished, or even executed. Because a large percentage of the scientists of the day were churchmen and quite religious people, this extreme position gradually changed and the church cautiously began to approve most of the scientific advances promulgated by those great scientists of that era, including Newton, Galileo, and Halley.

Now we can make a reasonable case that many of the leading scientists of our time are coming close to doing the same thing that the church did in the Middle Ages—that is, use our method (the scientific method) or we will not publish your papers or listen to your ideas. Their position seems to be, there is no other way to gain the truth than by using the scientific method. It is through a study of the natural world and only through a study of the natural world that we can learn the truth. The scientist operates with the assumption that we can eventually learn the secret of everything; in other words, there are no transcendental truths. Why do most scientists and educators now reduce the whole discussion of evolution down to Darwin's theories versus Creationism? Why are they so resistant to entertaining other possibilities? It seems that this is one area of science when any question of the status quo is met with derision and a great reluctance to examine the possibility that there may be other answers.

And of course, the age-old question—Should science have any moral or ethical restrictions? The development of nuclear energy with all its proven and theoretical negatives has presented us with some of the greatest challenges of our age. The development of the Internet, with instant communication among almost every human on the planet, with texting, distractions, the loss of interpersonal relationships, and a host of other negatives is threatening to change the social fabric of human beings all over the planet, in totally unexpected and unanticipated ways.

And of course, the great philosophical question—Where would we be if we could prove everything? All loss of wonder, awe, beauty, aesthetics, even hope, that essential ingredient for any sense of happiness. Would we lose these feelings if we completely understood them? Would we be better off? Would we be better off not knowing the full "objective" truth? The scientist is dedicated to finding the truth. What if, at the end of his research, a scientist "proved," using the accepted scientific methods and research principles, that humans are better off overall to live and act using both scientific and transcendental knowledge? Or the best course of action is in not seeking the ultimate truth in the physical world? Philosophers and moralists from the beginning—the forbidden fruit in the Garden of Eden—the early Greeks, have cautioned against the dangers of knowledge without transcendent considerations.

Why are so many scientists nonreligious or even antireligious? Why does the search for truth in the physical world turn people away from the transcendent things? And of course, on the other hand, many people after studying the natural world develop a strong belief in transcendent things. History is filled with the names of great and ordinary scientists who developed strong transcendent feelings because of their investigations of the natural world.

As scientists become more and more specialized from the very beginning of their learning and training, they have less and less time, if any, to learn anything about other subjects: philosophy, religion, or even other areas of science outside of their specialty. Their education is limited to their specific field and is lacking in exposure to the liberal arts. This lack of exposure to other areas of study may affect their understanding of their own specific subject.

As the body of human knowledge grows larger and larger, it becomes more and more difficult, if not impossible, for one individual to adequately understand a whole field of knowledge, and therefore, specialization is essential. In my area, the learning and practice of medicine, it is clearly impossible for one human being to learn all the facts and develop the technical skills required to cover all the various fields of medical practice. It is tempting to say that it would be more efficient in medicine and science generally to limit one's study at the very beginning of one's education to one specific field, and maybe we are nearing the point where this will be necessary to make progress. But what a loss that would be for the individuals involved to not know the joys and benefits of a general all-around or liberal arts education. And we clearly do not know whether reading Shakespeare makes a better doctor or reading history makes a better scientist.

And some thoughts about words, phrases, and semantics. A word or two can bring into consciousness a whole series of images and structures. If I say *house*, you immediately bring up a picture of a building with doors and windows and a roof and interior rooms, etc.

Very importantly, in our day, look at how we use common, everyday words and expressions to describe highly complicated technical machines or processes. An example is artificial intelligence, frequently identified simply by the acronym AI, or A.I. An experimenter may say that he has created a computer that is capable of performing certain tasks that are "intelligent"—that is, it performs tasks that approach the capacity of the human brain. Many laypeople, on hearing the phrase "artificial intelligence," assume the machine can do things entirely on its own and not merely reacting to the mechanics of the computer and its software. This is an oversimplification, but I think illustrates my point—that is, the reader or listener "hears" more than what is actually said and comes away misled.

And let us not even get started on how we have intermingled and intertwined our basic words with computer language and jargon. Suffice it to say that when we routinely refer to a massive collection of hardware and electronic equipment buried somewhere in the state of Utah as "the cloud," we may want to reconsider where we are headed.

This basic method of exchanging information is used by laypersons and scientists alike. We could not communicate without this process. But this process carries the danger of being misunderstood or misleading and the listener is left hearing or believing something that the speaker did not mean at all. And especially in science and objective endeavors, it is critical for the communications to be as clear and concise as humanly possible so that a miscommunication does not produce a series or a cascade of false information or false studies on the part of the reader or listener.

In summary, the fact that transcendental things can't be proved or have not been proved is not a valid reason to not consider them. It would be the same thing as refusing to consider any new scientific theory or idea because it is new and doesn't appear to be provable. A significant amount of science is accepted as valid, even though not provable because it works in most, if not all, situations.

Cannot the same criteria and reasoning be used when considering transcendental beliefs? They can't be proved, but they work in most, if not all, situations. Does this mean that we need to expand and add to the usual studies in our schools a formal section on the possibility of an Outside Force? We know how required studies of religion in public schools can be quickly co-opted by extremists and/or writers and thinkers who are not objective scientists. But it seems to me that if our point is to search for all truths, we must consider transcendent things. Does not a long historical record, dating over centuries, consistent with and describing transcendental things, offer a type of proof, or at least a valid reason to be included in a complete education?

As stated by Lecompte du Noüy in *Human Destiny*,

> *Indeed, man must beware more of scientific extrapolations than of moral ones, because his scientific experience has been much shorter than his psychological experience. New acts are frequently found in science which compel him to revise completely his former concepts. The history of science is made up of such revolutions: the atomic theory, the kinetic theory, the granular theories of electricity, energy, and light, radioactivity, relativity have successively transformed our point of view from top to bottom. The future of science is always at the mercy of new discoveries and new theories. The science of matter is not two hun-*

dred years old, while the science of man is over five thousand years old. Empirical psychology was highly advanced at the time of the third Egyptian dynasty, and great philosophers twenty-six hundred years ago displayed a knowledge of man which has not been surpassed, but only confirmed today. Therefore, it can be reasonably assumed that moral extrapolations are much safer than scientific ones, even though they cannot be expressed mathematically.

In a way, the scientists can claim an undefeatable position: what is unclear and unprovable today is relative, and with improved methods and additional knowledge, science will be able to prove everything and to disprove anything that today is excepted as a matter of faith, or unprovable. The scientist has *faith* that, in time, all things can one day be proved or disproved—science will one day fill in all the "gaps." The religious person has faith that there are things that are true that will never be proved.

Theories are an invaluable way of beginning any study of any subject and can be highly useful even when not proved. The molecular theory, which is still a theory, continues to be indescribably useful because it continues to answer almost all questions that are asked of it. This is not true of Darwin's theory. The Darwinian theory of evolution fails to answer so many important and basic questions, and that is why it continues to generate so much controversy. Part of the controversy, of course, is its bearing on religious faith, but my point is that it is poor science.

And finally, as Carl Sagan warned, all technological societies eventually destroy themselves. Is this just an inevitable result of progress? Can we do anything to avoid it?

And the bigger question—Is the failure of technological societies because of what technology produces, per se, or is it because those societies ignore the transcendent and come to believe that humans can progressively move toward a utopian existence through their own methods?

Chapter Fourteen
Artificial Intelligence

Technology is a useful servant but a dangerous master.
—Christian Lous Lange, Norwegian historian

Condensed: This chapter is presented as further support to the contention that the human brain is too complex to be a product of random forces. As human brains combine to try to develop machines that approach the ability of their own brains, we are reminded just how incredibly organized and capable our brains are.

*****

> *Artificial Intelligence (Dictionary.com)*: made by human skill; produced by humans. Capacity for learning, reasoning, understanding, and similar forms of mental activity; aptitude in grasping truths, relationships, facts, meanings, etc.

Artificial intelligence (AI)—right off the bat a misleading phrase. We are talking about a machine that has been programmed to do certain tasks and does not meet any of the criteria listed in the definition (except produced by humans). The machines should be called advanced electronic machines mimicking rudimentary human brain function. Using the phrase "artificial intelligence" leads the reader to think that in some way, these machines compare to the ability of the human brain. We must be very careful using words and

ARTIFICIAL INTELLIGENCE

phrases that may lead us down paths of thinking and thought processes that we never intended. To suggest that a mechanical machine is remotely similar to a human brain does not elevate the concept of artificial intelligence as much as it lowers the dignity of humans.

Any type of artificial intelligence created by human beings (that is, by human brains) has two be made out of the materials at hand—that is, the elements, compounds, etc. of the planet Earth. For all our dreams and fantasies, let us not forget that function always depends on structure. The billions and billions of cells and connections and interconnections in the human brain cannot be replaced by a physical structure anywhere near the same size of the human brain without having an unimaginable superior operating system and/or logic. At the present time, this is not in the foreseeable future.

To claim that the human brain can create a device that is equal to it, or even more capable, is ludicrous. The problem of originality—that is, the problem of creating something that exceeds the physical capability of the device created—seems insurmountable. Again we are back to the problem: Can the total be greater than the sum of its parts? Add to this the limitations of the technology of the present era—that is, the present-day computer and computer languages all based on Boolean logic (on or off electrical switches and algorithms) and digital technology. The size of such a computer approaching the capability of the average human brain would have to be the size of the moon.

Most of the experts in this field agree that the development of an artificial intelligence anywhere near the capabilities of the human brain would require an entirely new type of hardware, software, operating systems, logic, etc., none of which are on the horizon in the year 2019. In other words, we would need a whole new technology to replace our digital systems.

The history of mankind is the history of technological advances, but many, if not most, have not been anticipated, and a type of intelligence far more capable than the human brain of 2019 may be developed. But our discussion at present is to analyze the information we have at hand and to develop conclusions that are as near to the truth as we are capable of doing.

I had an interesting talk with an atheist (secular humanist) who believes that the human body, with its brain, is simply an organism developed through evolutionary processes over a long period of time and in essence has no inherent value. *Homo sapiens* has developed systems of moral and ethics simply because they are the best results of adaptability and ensuring survival of the species. There is nothing outside of the physical structure of the body, just as there is nothing outside of animals or even plants or minerals.

I asked him the following question: If human beings created an entity similar to human beings, with a brain similar to the human brain and body, being constructed of the hard elements of the planet—that is, plastic, copper wires, batteries, rubber, and a multitude of other nonprotein materials—would you consider this entity equal to *Homo sapiens*? His answer was that he could see no difference. Human beings were "created" by impersonal forces, and this new entity was created by human beings. No difference. No innate value. Just like human beings, no emotions, no morals, no ethics, and no restrictions on behaviors that are not related to survival or preventing pain, or causing destruction or discomfort on fellow creatures. One created out of proteins and the other created out of different, simpler materials. But basically, in the overall scheme of the universe, no difference.

This leads to some interesting and challenging questions:

- Since the AI has no value, could human beings use it as a slave, "kill" it, abuse it, etc., without reservation?
- The humanists would say that because the new entity is equal to humans, and since they do not believe that human beings should be treated as slaves, abused, or killed, then humans could not treat these creations differently. They could not be unplugged nor have their batteries removed.
- How about if the artificial intelligence becomes "better" than humans—that is, more efficient, needing less maintenance, using less resources and better for the environment and long-term occupation of the

planet, more moral and ethical with better relations with each other, no wars or horrors—should they destroy or do away with *Homo sapiens* as inferior and destructive and not deserving a space on the planet?
- Would AI have any biases? If all biases and prejudices are programmed out, would that automatically improve the superiority of its "intelligence"?
- At our present level of understanding of the human brain, scientists and philosophers have basically no understanding of the concept of "consciousness"—that is, the human brain having an awareness of its own existence. This being true, the AI could not be programmed to have "consciousness" since we ourselves do not understand consciousness. Without this vital ability, how could we ever claim that the AI approaches human intelligence or ability?
- Given our present state of technological ability, is it conceivable that we could add wisdom, that ability that human beings use to think and act using knowledge, experience, understanding, common sense, and insight, to our artificial intelligence machines? And the other human attributes such as compassion, and virtues such as ethics and benevolence? Could we add another human trait, sentience—that is, feeling or sensation as distinguished from perception and thought?
- Without a sense of consciousness, could the AI be programmed to include the necessary component of self-survival, an essential ingredient that has sustained human life?
- Would suicide and euthanasia be programmed in the AI, or would it pose no problem for the AI?
- And lastly—a way-out question—is what we call artificial intelligence the next step in the evolution of *Homo sapiens*? It meets the criteria for advancement by mutations and survival of the fittest. Is it a step

too far to think that artificial intelligence, made of silicone, is being influenced by what I have referred to as the Outside Force? That is, has Outside Force chosen to use silicone, copper, batteries, etc. to advance his creation?

And finally, and most importantly, is the question, Would we, or could we, ever make the claim that our creation is "alive"? Can *Homo sapiens* ever cross that line that differentiates an entity being "dead" or "alive"? Man's age-old dream has been to create life. Would our efforts result in a Frankenstein monster of Mary Shelley's novel, or would we humans with all our limitations and vices create a better individual? Is it possible for our brains to conceive of or understand the ramifications of such an effort?

Chapter Fifteen
Analysis

The aim of this book is to make the case for a transcendent power or being—that is, that there is a specific entity residing beyond the ability of human beings to understand using purely objective studies, e.g., science. The scientist, theologian, philosopher, and all other humans are limited by the ability of the human brain; we simply cannot comprehend or understand everything in the natural world or in the mental, cerebral world where we humans reside. Every effort we make simply reaches a point we can go no further.

We described with words and pictures that tiny speck of the universe where human beings live, look out into the vast universe, and look down into the incredibly small world of the microcosm.

We presented several chapters highlighting the scientific approach to examining and understanding the complexities of the physical world. We looked at the strengths and weaknesses of the scientific approach, how this approach has added so much to the comfort and progress of mankind, but has its limitations and outright dangers.

We specifically presented the strengths and weaknesses of Charles Darwin's theories regarding the evolution of human beings. We presented the case that accepting evolution of human beings purely through random mutations and survival of the fittest reduces humanity to nothing more than all the other animals, even vegetation or rocks.

We presented a chapter on the status of the new science of artificial intelligence, an outgrowth of the development of computers in the digital world—the idea that human beings can create an entity, not just with the physical abilities of the human body, but with mental and emotional capabilities equal to the human brain.

We started with this basic assumption: there is a reason for everything. There has to be some force or pressure to cause a change. The scientists use chance and probability to help them draw conclusions and develop theories, which are then tested and found to be useful or not. They are not concerned with a basic reason or primary cause. The philosopher and theist seek for a reason or a cause for the existence of the universe, and the reason for any subsequent changes. The history of human's stay on this Earth is filled with multiple answers and theories to try to satisfy this longing.

There are five generally accepted causes proposed for the existence of the universe and the place of human beings in that universe.

- The big bang was the beginning of the universe, and thereafter every single event in the universe from that moment to the present was fixed and inviolable and responds only to natural laws—that is, determinism, without any reference to a prime mover or first cause.
- A Prime Mover created the universe as we know it by creating the big bang and then standing neutrally aside as it evolves—that is, the "wind-up clock" explanation followed by determinism.
- A Prime Mover created the universe using the mechanism of the big bang, allowing the universe to evolve, generally constrained by the natural laws imposed by the Prime Mover, but from time to time interjecting its will into the process to effect certain changes, moving the universe toward the Prime Mover's ultimate goal. This explanation also includes the ability of human beings to make free decisions, not determined or controlled by the natural law.

ANALYSIS

- A Creator created the universe, in the relatively recent past, in a short period of time, complete and finished, without any evolutionary changes. This Creator is actively involved in the history of the universe, from time to time changing or contradicting the natural law, and has given to his creation, human beings, free will to make original decisions and changes. This position is generally referred to as Creationism.

A computer "freezes," a glitch appears, any number of problems or "changes" happen. The technicians may or may not be able to find a specific reason for the change, but they know that something happened electronically or mechanically to cause the change. They do not claim it was random. A programming error that finally appeared, a static electrical discharge, a failure of a part of the hardware; something caused the change. But the technician and all who use the computers know beyond a doubt that humans created this machine, that the machine undergoes changes for good or bad, and that there is always a reason for these changes.

Any objective effort to understand the place of human beings in the overall scheme of the universe must include a study of the science, especially as it pertains to the evolution of the human body with its brain. We must study both the natural world and the world of the transcendent to fully understand all the multiplicity of things that enter our brains through our senses. How we understand these things affects our thinking about our origin and development as human beings, especially our value and dignity. It's not the same as pure science, and as in the pursuit of truth in the natural sciences, we must be totally committed to the quest for truth in the study of our origins and transcendent beliefs.

We presented the case, even without considering primary causes or looking back before the big bang, that Darwin's theories provide an inadequate explanation in scientific terms for the causes of human evolution. Science works with things that can be proved. Science admits that there are things that cannot be proved, but it also accepts many things that cannot be proved by faith and continues to work with these things believing one day they will be proven to be true. They often fill in the gaps in science just as theologians fill in the gaps in religious faith.

And one of the most important points made is that the scientist, as well as all of us, must realize that the human brain is limited and there is the possibility of a primary cause which is beyond our brain's ability to understand. The chapter on the human brain went into some detail to present the human brain as being very complex, too complex to have evolved simply by random changes with no guidance.

The incredible complexity of our planet, including life and inanimate things, and what we have learned so far about the universe, makes us stand in awe, humbled by our knowledge and our obvious lack of understanding. Those in the future will certainly look back on our era as an era of great advancement but also, as always in human history, containing many illusions and false assumptions and even laughable theories.

What now do we tell our children or those who come after? The problem with teaching our children, and those who have not had the opportunity to be fully informed about these issues, is that they have to have some guidance.

There are many topics that are incredibly difficult, if not impossible, to learn without a teacher or mentor. The teacher, as well as the scientist, should not teach values, or especially introduce their own values into the teaching of science, or any subject for that matter. This is especially true with teaching evolution, which, as we pointed out earlier, affects our basic understanding of human life and therefore lends itself to the impossible task of not interjecting values or denying the possibility of transcendent things. Without some form of guidance in the teaching of this scientific subject, the student can be led down any number of paths, including paths that do not result in their understanding the truth, which is, of course, the main goal of science as well as those studying transcendent things.

We all need to understand that if one accepts without question the generally accepted method of teaching of evolution, it has the potential to affect one's feeling about the dignity of human life and the full enjoyment of life. Many of the most enjoyable and meaningful things in life can become meaningless. Obviously, for some, this is not true. Almost every human being who believes in traditional Darwinian evolution and determinism appears to have the same feelings and emotions about life as those who do not believe in Darwinian evolution and determinism. Why? At a minimum, this certainly suggests that there is at least a basic impulse that resides in the evolved human brain that has kept transcendental influences alive and enduring through the centuries.

And finally, give Darwin his due. He was the main motivator for the study of evolution, and this has clearly helped us in our understanding of the physical world. His two major books and most of his other works still stand as monuments to scientific inquiry. We must continue to look for evidence of the evolutionary processes and develop new theories for the causes, realizing that our understanding is far from complete and that the truth always includes the possibility of transcendent things.

There is no question that we cannot figure out or understand all these points. The scientists fully realize that many of their findings are statistical and freely admit that many of their theories and understandings can never be proved hundred-percent. They are comfortable knowing that although there is a faint possibility of error, for all practical and pragmatic purposes, a theory that works is acceptable. But the point remains that over the course of human history, the vast majority of human brains, from the poorly to the highly organized, have had transcendent feelings and unshakable beliefs. Cannot these beliefs be accepted by the scientific mind as usable and pragmatic, since, statistically, the probabilities are the same: not completely or totally provable, but in every way better?

Chapter Sixteen

Summary

In the observable world causes are found to be ordered in series; we never observe, nor ever could, something causing itself, for this would mean it preceded itself, and this is not possible. Such a series of causes must, however, stop somewhere; for in it, an earlier member causes an intermediate and the intermediate a last (whether the intermediate is one or many). Now if you eliminate a cause you also eliminate its effects, so that you cannot have a last cause, nor an immediate one, unless you have a first. Given therefore no stop in the series of causes, and hence no first cause, there would be no intermediate causes either, and no last effect and this would be an open mistake. One is therefore forced to suppose some first cause, to which everyone gives the name "God."

—Thomas Aquinas

In the beginning there was the unknown. Then came the big bang. Then came the universe. Then came the elements. Then came the planet Earth. Then came the big ooze. Then came cells and tissues and organs. Then came life. Then came vegetation. Then came animals and prehistoric humans. Then came *Homo sapiens*. Now we have modern humans with their brains.

How did modern humans get from there to here? That is the question. This question has interested, fascinated, and, indeed, plagued us humans since we developed consciousness. Our theories and understanding range from total determinism and "chance" to a simple creation story by some Outside Force that we cannot fully comprehend or prove. It is not an exaggeration to say that this question, the search for an answer, and the myriad of conclusions reached through the ages is one of the most important motivations for the behavior of human beings on this planet. Therefore, it is extremely important in the education of our youth, as well as a continual search for truth for adults, that we make sure we all understand that there are limitations to science and objectivity and that the persistent, unquenchable thirst for the transcendent has been a part of the character of most humans since *Homo sapiens* became conscious of its existence. We must never succumb to the illusion that if it cannot be proved, then it does not exist.

To support this contention, we started with the big bang, which is where our science starts, and presented a series of chapters tracing the evolutionary processes resulting in the most advanced structure of those processes, the human brain. The pros and cons, strengths and weaknesses of the commonly accepted maxims and theories were presented. The point was made that science and the scientific method is clearly responsible for much of the progress and advancement of the human species, but that this method is not infallible, and history is full of examples of errors and contradictory conclusions.

The chapter on the human brain pointed out its inherited capabilities and how it accumulates knowledge and new data, how human beings interpret and understand the natural world, how there is a possibility that there are other inputs into the human brain that the brain itself is not conscious of. We noted that the brain is limited and that there is more to reality than the human brain understands, or specifically, is capable of understanding. We reviewed chance, probability, randomness, and determinism as answers to our understanding of the natural world. We then discussed consciousness and free will as abilities specifically limited to a function of the human brain. We then discussed Charles Darwin's theory of evolution and its peculiar position as a scientific study with great philosophical and theological implications—that is, the value of human beings.

There was a chapter on the present status of what is called "artificial intelligence" and how this new science presents a new challenge as to the value and dignity of human beings. And finally, we discussed the fascinating question: Do myths, properly interpreted, present truths, although objectively and scientifically they never existed?

As the writer Malcolm Muggeridge stated, *"If and when we know the final truth about human life, we shall find that the legends, or what pass for legends, are far nearer the truth than what passes for fact or science or history."*

The main strategy was to back up the claim that there is a reason for everything. Modern science, totally committed to those things that can be proved, goes back no further than the big bang for causes. The scientist's efforts are completely directed toward finding the "what" and the "how" of the natural world. In their professional lives, scientists are not interested in the "why." This is as it should be.

But ending the search with the what and how and not pursuing the why leaves a desolate, empty feeling. From the beginning of consciousness, human beings with few exceptions have been obsessed with the why of things. The scientist, in his work as an investigator and researcher, is interested in causes in so far as an explanation for the specific project he is working on. He takes it back only so far as to give him a "reasonable" explanation. This quite frequently ends with an explanation that basically says, "The answer meets the standards of probability, or is due to 'chance.'" The scientist understands that this is not completely satisfactory, but it works most of the time and enables him to accurately predict the outcome of the changes in the natural world. But the scientists, of course, outside of their laboratories and in their private hours, can have their convictions about an Outside Force or realities beyond proof. Indeed, many, if not a large majority, of the brightest and most studious have responded to this inborn quest to explain the ultimate or primary cause.

The scientific approach has been incredibly productive as far as material things are concerned. But we must understand that there is much more to reality than can be proved or studied by scientific methods and objective investigation—that is, there is a possibility that there are things beyond our ability to study, indeed, beyond the capability of the human brain to comprehend. That is why "brain limit" is such an important part of the understanding of this treatise. It's not that we don't understand; it's that we aren't capable of completely understanding these issues.

*****

What is the case for the transcendent? The first, foremost, and most compelling reason is that persistent feeling, through all of recorded history and in all cultures and almost every human being, that there is something beyond our ability to fully comprehend or understand. Even the prehistoric cavemen left us pictures and drawings on the walls of their caves that present incontrovertible evidence of feelings and urgings beyond the daily struggles of life. Going back to the time of the early Greek philosophers, almost every human being has believed that there is, in Aristotle's phrase, "a Prime Mover," or some mysterious cause for the creation and development of the planet and human beings. There is embedded in the very matrix of the human brain a desire to understand and seek an answer to this mystery. This feeling simply will not leave us alone.

To use an analogy that has been suggested by some, although many may consider it crude, let us compare the overwhelming desire to include the transcendent in our lives to the sex drive. These two feelings have been persistent, strong to the point of risking life and limb for satisfaction, pushing individuals to the extremes of good and evil, necessarily including two parties and a relationship, and feelings rarely denied by any individual. Everyone agrees that the sex drive is an essential part of the basic makeup of human beings, and there is no continuity or survival without this drive. In a sense, the persistent drive to include a relationship with a transcendent party matches the intensity of the sex drive.

Secondly, history is filled with the names of its greatest geniuses—Isaac Newton, Galileo Galilei, Nicolas Copernicus, Blaise Pascal, Thomas Aquinas, and a host of others—who believed in and were comfortable with the belief in a transcendent power. Notably, many of these enlightened scientists also felt strongly that this transcendent power was personal and influenced the lives of human beings on the planet Earth. The claim that they were shallow in their understanding of things outside of the natural world, or that they were simply the product of childhood indoctrination or cultural or social influences, while at the same time producing some of the most objective and scientific studies in human history is itself not objective. This claim challenges not only their integrity but their intelligence.

Albert Einstein, the greatest scientist of the modern era, made many statements about his belief, or nonbelief, in a reality beyond reason and the scientific approach. The following paragraph makes it clear that Dr. Einstein realized that there was some power, some force that operates outside of all the physical universe as we understand it.

You will hardly find one among the profounder sort of scientific minds without a religious feeling of his own. But it is different from the religiosity of the naive man. For the latter, God is a being from whose care one hopes to benefit and whose punishment one fears, a sublimation of a feeling similar to that of a child for its father, a being to whom one stands, so to speak, in a personal relation, however deeply it may be tinged with awe. But the scientist is possessed by the sense of universal causation. The future to him is every whit as necessary and determined as the past. There is nothing divine about morality; it is a purely human affair. His religious feeling takes the form of a rapturous amazement at the harmony of natural law, which reveals an intelligence of such superiority that, compared with it, all the systematic thinking and acting of human beings is an utterly insignificant reflection. This feeling is the guiding principle of his life and work, in so far as he succeeds in keeping himself from the shackles of selfish desire. It is beyond question closely akin to that which has possessed the religious geniuses of all ages.

Thirdly, every scientific effort, with few exceptions, ends with an incomplete understanding: that is, this is useful information and can be used with good and predictable results, but we still do not understand the primary cause.

And fourthly, the belief in a transcendent force has benefited mankind in general and been a major force in the evolution of humanity—that is, it has been not chance and random mutations but an Outside Force pushing mankind toward its destiny.

SUMMARY

The problem is not so much believing in, or accepting the possibility of, an Outside Force, but in analyzing and understanding the myriad of responses by individual human beings and cultures to this possibility. Statues, idols, weather patterns, worshipping political or military leaders, and a host of other practices are part of the multitude of responses and actions developed by human beings toward this feeling that there is something more to human existence.

Among the many questions that have been asked through the ages are, "Is this force personal?" "Does this force interfere with the natural law or human behavior or decision-making," or "Did the force 'wind up the clock' and became only an observer?" "Is the force 'good' or 'bad' or indifferent?" If an individual does not have a belief in an Outside Force, why does that individual feel compelled to behave morally and ethically? Is this simply a Darwinian trait predetermined by natural forces? If so, what part does it play in survival of the fittest? Is the Outside Force influencing this individual outside of his consciousness?

Chapter Seventeen

Epilogue

In order to understand, you must believe.
— St. Augustine

Every human being, with rare exceptions, experiences exhilarating and "otherworldly" feelings that cannot be defined or analyzed objectively. We study poetry, music, literature, art, and all aesthetic experiences and human endeavors that uplift mankind and make human life more enjoyable. We have no reservations about offering, or even requiring the teaching of these subjects in our public institutions. But in so far as including the long history of transcendent beliefs in the lives of human beings in our educational courses, we dismiss them and treat them as if they never existed. Go to any library, public or private, and you'll note that a large percentage of the books are books on transcendent things. But in the classrooms, this particular area of human history is the elephant in the room: obviously there but ignored.

The overriding motivation for this treatise was to help those who react negatively to the study of transcendent things to try to realize that even though they cannot be studied using the scientific method, they can be studied objectively with an open mind. The author hopes he has succeeded in stimulating those who have excluded all transcendent possibilities to think again about the value and exciting possibilities of adding transcendent things to their body of knowledge just as much as learning about the physical world.

EPILOGUE

Anyone who has studied or is aware of the incredible differences of beliefs, opinions, and feelings about what is called an "Outside Force" or "the transcendent" will realize that the individuals holding these beliefs cover a broad spectrum. The spectrum ranges from indifference to worship, from so-so to awe, from completely impersonal to as close a relationship as parent to child.

Simply stated, is the Outside Force or the transcendent personal or impersonal? The impersonal is the position expressed by Albert Einstein and others: the belief that there is a reality beyond our ability to understand, and this belief creates a sense of awe in us and makes us realize how small we are in the universe, but is neutral in its relationship to human beings.

The other position, the personal position, is the belief that there is a force operating outside of the human realm—a force that cannot be seen or studied, but exists and interacts with human beings, making certain demands of humans and promising rewards on this Earth, and for some a belief in an afterlife and eternal paradise. And a small minority of humans recognizes the existence of an Outside Force, but the force does not deserve respect or worship and is treated indifferently.

The following is a summary of the author's thoughts on an Outside Force.

Let's assume the possibility of an Outside Force. This is my transcendent force—that is, the influence that acts outside of the natural law as we understand it. This force acts on the universe in the following ways:

- From time to time it changes the natural law (miracles).
- From time to time it changes the thoughts of human brains.
- From time to time it changes the chromosome pattern (basically the same as changing the natural law).

This obviously produces problems for religious thinkers as well as the pure scientist or pragmatist. It seems absurd to believe that this force would influence different brains (humans) in different ways, resulting in the creation of multiple belief systems, with their inherent differences and conflicts.

The author's answers…

- Stick with what you believe. This almost always requires some degree of cognitive dissonance—that is, holding two or more conflicting beliefs at one time.
- Realize that the incompatibilities cannot be made compatible, and don't fight with windmills.
- We must be very careful in how we perceive the influence of an Outside Force. If the Outside Force basically dictates our behavior and changes, then this modifies our free will, negating or partially negating that thing that makes us human—e.g., our ability to choose values.
- There can never be any "proof" of transcendent things. When the scientists pushed the button at Alamogordo, New Mexico, in 1945, and the bomb went off, there was no doubt that the scientists had "proved" a great truth about the physical world. When one sits in a magnificent cathedral listening to a great orchestra with the voices of the chorus and the choir performing Handel's *Messiah*, while this does not "prove" the existence of transcendental things, I would submit to the world that this is similar to the correlation and causation that the scientists experienced in the New Mexico desert.

*****

After finishing this project, I realized that I had ended where I began—that the most profound emotions are the things we cannot fully understand; that those things we cannot fully understand are the driving forces that lead human beings to create the things that

EPILOGUE

uplift us and produce the things that are best in life; that those who don't accept the transcendental things miss out on the best of life; to know that there are things our brains are incapable of understanding really exists; that things that are beyond objective investigation "manifest themselves as the highest wisdom and most radiant beauty" (Einstein's words); that this knowledge and feeling is the very essence of what makes us unique in this universe…

And confirmed by Werner Heisenberg:

> *In the history of science, ever since the famous trial of Galileo, it has repeatedly been claimed that scientific truth cannot be reconciled with the religious interpretation of the world. Although I am now convinced that scientific truth is unassailable in its own field, I have never found it possible to dismiss the content of religious thinking as simply part of an outmoded phase in the consciousness of mankind, a part we shall have to give up from now on. Thus in the course of my life I have repeatedly been compelled to ponder on the relationship of these two regions of thought, for I have never been able to doubt the reality of that to which they point.*
>
> *Where no guiding ideals are left to point the way, the scale of values disappears and with it the meaning of our deeds and sufferings and at the end can lie only negation and despair. Religion is therefore the foundation of ethics, and ethics the presupposition of life.*

And completed by Lecompte du Noüy:

> *Let everyman remember that the destiny of mankind is incomparable and that it depends greatly on his will to collaborate in the transcendent task. Let him remember that the Law is, and always has been, to struggle and that the fight has lost nothing*

of its violence by being transposed from the material onto the spiritual plane; let him remember that his own dignity, his nobility as a human being, must emerge from his efforts to liberate himself from his bondage and to obey his deepest aspirations. And let him above all never forget that the divine spark is in him, in him alone, and that he is free to disregard it, to kill it, or to come closer to God by showing his eagerness to work with Him, and for Him.

THE END.

For now we see through a glass, darkly; then we shall see face-to-face: now we know in part; then shall we know even as we are also known. (St. Paul)

About the Author

Dr. C. Thomas Cook is a retired medical doctor living in Charleston, South Carolina. He is a graduate of Furman University and the Medical University of South Carolina. After completing his medical training, Dr. Cook served in the United States Air Force in Vietnam in the middle of the Vietnam War years, and with his wife, a registered nurse, returned to Vietnam as a civilian doctor working with Project Concern, a medical relief organization with two hospitals in Vietnam. During these years, Dr. Cook worked with the indigenous populations and was exposed to the religions of the Far East, mostly Buddhism.

Dr. Cook's experiences include undergraduate studies in religion, philosophy, and English literature. In medical school, students learn about the conception and development of the human embryo through its various stages of growth, and finally delivering the final product as a newborn child. During the years of his practice and training, Dr. Cook delivered multiple babies, performed over 120 autopsies, and over 32 years practicing Emergency Medicine experienced an incredible array of human illnesses, complicated physical and physiological reactions to diseases, and the many psychological and psychiatric conditions that affect the human brain.

CPSIA information can be obtained
at www.ICGtesting.com
Printed in the USA
LVHW070211290520
656886LV00028B/1747